Springer Proceedings in Earth and Environmental Sciences

Series editor

Natalia S. Bezaeva, The Moscow Area, Russia

The series Springer Proceedings in Earth and Environmental Sciences publishes proceedings from scholarly meetings and workshops on all topics related to Environmental and Earth Sciences and related sciences. This series constitutes a comprehensive up-to-date source of reference on a field or subfield of relevance in Earth and Environmental Sciences. In addition to an overall evaluation of the interest, scientific quality, and timeliness of each proposal at the hands of the publisher, individual contributions are all refereed to the high quality standards of leading journals in the field. Thus, this series provides the research community with well-edited, authoritative reports on developments in the most exciting areas of environmental sciences, earth sciences and related fields.

More information about this series at http://www.springer.com/series/16067

Igor Bychkov · Victor Voronin
Editors

Information Technologies in the Research of Biodiversity

Proceedings of the International Conference "Information Technologies in the Research of Biodiversity", 11–14 September, 2018, Irkutsk, Russia

 Springer

Editors
Igor Bychkov
Matrosov Institute for System Dynamics
and Control Theory
Siberian Branch of Russian Academy
of Sciences
Irkutsk, Russia

Victor Voronin
Siberian Institute of Plant Physiology
and Biochemistry
Siberian Branch of Russian Academy
of Sciences
Irkutsk, Russia

ISSN 2524-342X   ISSN 2524-3438   (electronic)
Springer Proceedings in Earth and Environmental Sciences
ISBN 978-3-030-11719-1   ISBN 978-3-030-11720-7   (eBook)
https://doi.org/10.1007/978-3-030-11720-7

Library of Congress Control Number: 2018967733

This Springer imprint is published by the registered company Springer Nature Switzerland AG
The registered company address is: Gewerbestrasse 11, 6330 Cham, Switzerland

Preface

This volume contains the proceedings of the First International Conference "Information Technologies in the Research of Biodiversity" that was held from 11 to 14 September 2018 in Irkutsk, Russia. The conference includes plenary and sectional reports, round tables and master classes on the publication of biodiversity data for use in basic and applied research to provide you with principles for working with existing biodiversity information systems.

Chapters in this book cover areas of interaction of knowledge on biodiversity and information technologies. Integration of modern information technologies and classical methods of biodiversity research enables reaching new levels of knowledge acquisition. New demands on researchers require coordination, standardization and unification of data and methods.

The main topics include methods, models, software systems for the analysis of biodiversity data; global data portals; information and analytical systems on biodiversity; application of remote methods in vegetation mapping; and theoretical fundaments and organization technologies of the information and telecommunications infrastructures.

The conference was organized by Matrosov Institute for System Dynamics and Control Theory SB RAS in cooperation with Siberian Institute of Plant Physiology and Biochemistry SB RAS, V.B. Sochava Institute of Geography SB RAS, Institute of Mathematical Problems of Biology RAS—the Branch of the Keldysh Institute of Applied Mathematics RAS. The conference was held with the support of the GBIF —Global Biodiversity Information Facility and Department of Nanotechnology and Information Technologies RAS.

The organizers would like to thank all the authors for their participation, Technical Program Committee, Editorial Board and associated reviewers for their good work and important contribution to the conference.

September 2018

Igor Bychkov
Victor Voronin

Organization

General Chair

Igor Bychkov   Matrosov Institute for System Dynamics and
Control Theory of Siberian Branch of Russian
Academy of Sciences, Russia

Program Chairs

Igor Bychkov   Matrosov Institute for System Dynamics and
Control Theory of Siberian Branch of Russian
Academy of Sciences, Russia

Victor Voronin   Siberian Institute of Plant Physiology
and Biochemistry of Siberian Branch of Russian
Academy of Sciences, Russia

Program Committee

Anne-Sophie Archambeau   University Pierre and Marie Curie GBIF, France

Oleg Borodin   State Scientific and Production Association "State
Research and Development Center for Biological
Resources of the National Academy of Sciences
of Belarus", Minsk, Republic of Belarus

Ingimar Erlingsson   Swedish Museum of Natural History, Stockholm,
Sweden

Alexei Hmelnov   Matrosov Institute for System Dynamics and
Control Theory of Siberian Branch of Russian
Academy of Sciences, Russia

Mikhail Kalyakin	Lomonosov Moscow State University, Moscow, Russia
Ma Keping	Institute of Botany, Chinese Academy of Sciences, Beijing, China
Dmitry Schigel	Global Biodiversity Information Facility Secretariat, Copenhagen, Denmark
Nadezhda Konstantinova	PABGI KSC RAS, Apatity, Russia
Svetlana Kubrak	Institute of Genetics and Cytology of the National Academy of Sciences of Belarus, Minsk, Republic of Belarus
Otso Ovaskainen	University of Helsinki, Helsinki, Finland
Varos Petrosyan	Severtsov IPEE RAS, Moscow, Russia
Gennadiy Ruzhnikov	Matrosov Institute for System Dynamics and Control Theory of Siberian Branch of Russian Academy of Sciences, Russia
Alexey Seregin	Lomonosov Moscow State University, Moscow, Russia
Alexey Shekhovtsov	V.B. Sochava Institute of Geography SB RAS, Russia
Mikhail Ustinin	IMPB RAS—the branch of Keldysh IAM RAS, Pushchino, Russia
Igor Vladimirov	V.B. Sochava Institute of Geography SB RAS, Russia
Andrey Zverev	Tomsk State University, Tomsk, Russia

Contents

**The Use of NDVI for the Analysis of the Effect of Drought
on Vegetation Productivity in the Pre-Urals Steppe Area
Where a Population of the Przewalski Horse Equus Ferus
Przewalskii Polj., 1881 Had Been Established** 1
Nikolay I. Fedorov, Tatyana L. Zharkikh, Oksana I. Mikhailenko,
and Rafilya T. Bakirova

**Regional Information and Analytical System on Insect Biodiversity
of the Baikal Region** 8
Igor A. Antonov

**Use of Remote Sensing Data and GIS Technologies for Monitoring
Stocks of Medicinal Plants: Problems and Prospects** 14
N. B. Fadeev, T. N. Skrypitsyna, V. M. Kurkov, and N. I. Sidelnikov

**Digital Herbarium Collections of the Central Siberian Botanical
Garden SB RAS, Novosibirsk, Russia** 22
Nataliya Kovtonyuk, Irina Han, and Evgeniya Gatilova

**Taxonomic and Phytogeograpical Databases in Systematics
of the Flowering Plant Family Umbelliferae/Apiaceae** 28
Michael G. Pimenov, Michael V. Leonov, and Tatiana A. Ostroumova

**The Fractal Model of the Microorganism's Frequencies Spectrum
for Determining the Diversity of the Biochemical Processes
in Soil** 37
N. I. Vorobyov, V. N. Pishchik, Y. V. Pukhalsky, O. V. Sviridova,
S. V. Zhemyakin, and A. M. Semenov

Actualization of Herbarium Labels Information 42
Olga Pisarenko, Igor Artemov, Sergey Kazanovsky,
and Ekaterina Prelovskaya

**Ground Surveys Versus UAV Photography: The Comparison of Two
Tree Crown Mapping Techniques** 48
Maxim Shashkov, Natalya Ivanova, Vladimir Shanin, and Pavel Grabarnik

**Creation of Information Retrieval System on the Unique Research
Collections of the Zoological Institute RAS** 57
Oleg Pugachev, Natalia Ananjeva, Sergey Sinev, Leonid Voyta,
Roman Khalikov, Andrey Lobanov, and Igor Smirnov

**The Use of Mathematical Methods in Analysis of Antibioticresistans
of Microorganisms of Lake Baikal** 66
E. V. Verkhozina, A. S. Safarov, V. A. Verkhozina, and U. S. Bukin

**Mapping of Model Estimates of Phytoplankton Biomass
from Remote Sensing Data** 73
Svetlana Ya Pak and Alexander I. Abakumov

**About the Project of the Web GIS "Electronic Atlas of Bryophytes
of the Republic of Bashkortostan"** 80
T. U. Biktashev, N. I. Fedorov, and E. Z. Baisheva

Bioclimatic Data Optimization for Spatial Distribution Models 86
Mikhail Orlov and Alexander Sheludkov

**Forest Resources of the Baikal Region: Vegetation Dynamics
Under Anthropogenic Use** 96
Anastasia K. Popova, Evgeny A. Cherkasin, and Igor N. Vladimirov

**DNA Barcoding of *Waldsteinia* Willd. (Rosaceae) Species
Based on ITS and *trnH-psbA* Nucleotide Sequences** 107
Marina Protopopova, Vasiliy Pavlichenko, Aleksander Gnutikov,
and Victor Chepinoga

**Technology of Information and Analytical Support
for Interdisciplinary Environmental Studies in the Baikal
Region** 116
Igor V. Bychkov, Gennady M. Ruzhnikov, Roman K. Fedorov,
Yurii V. Avramenko, Alexander S. Shumilov, Alexei O. Shigarov,
Alla V. Verhozina, Natalia V. Emelyanova, and Andrei A. Sorokovoi

**Geoinformational Web-System for the Analysis of the Expansion
of the Baikal Crustaceans of the Yenisei River** 125
A. V. Andrianova and O. E. Yakubailik

**Some Problems of Regional Reference Plots System for Ground
Support of Remote Sensing Materials Processing** 131
Alina Bavrina, Anna Denisova, Lyudmila Kavelenova, Eugeny Korchikov,
Oksana Kuzovenko, Nataly Prokhorova, Darya Terentyeva,
and Victor Fedoseev

Late Cenozoic Lagomorphs Diversity in Eurasia 144
M. A. Erbajeva

An Instrumental Environment for Metagenomic Analysis 151
Evgeny Cherkashin, Alexey Shigarov, Fedor Malkov,
and Alexey Morozov

**Experience in the Use of GIS Tools in Plant Systematics
and Conservation** 159
M. Olonova, D. Feoktistov, and T. Vysokikh

Diatom Analysis Using SOQL Language Interpreter 169
Y. V. Avramenko, R. K. Fedorov, and A. D. Firsova

Database of Barguzinsky Reserve 174
Evgeniya Bukharova, Alexander Ananin, and Tatiana Ananina

Development of Complex GIS Monitoring of the Angara River 181
Andrey S. Gachenko and Alexei E. Hmelnov

**Current State and Rational Use of Landscapes in the Border Area
of Mongolia and Russia** 187
Alexey I. Shekhovtsov, Irina A. Belozertseva, Igor N. Vladimirov,
and Darya N. Lopatina

**Environmental Aspects of Urbanized Territories in the
Baikal Region** 193
Olga V. Gagarinova, Andrey Sorokovoy, Irina A. Belozertseva,
Natalia V. Emelyanova, and Roman Fedorov

Author Index 201

The Use of NDVI for the Analysis of the Effect of Drought on Vegetation Productivity in the Pre-Urals Steppe Area Where a Population of the Przewalski Horse Equus Ferus Przewalskii Polj., 1881 Had Been Established

Nikolay I. Fedorov[1]([✉]) [iD], Tatyana L. Zharkikh[2] [iD],
Oksana I. Mikhailenko[3] [iD], and Rafilya T. Bakirova[2] [iD]

[1] Ufa Institute of Biology, UFRC RAS, Prospect Oktyabrya, 69, Ufa, Russia
fedorov@anrb.ru
[2] The Federal Government Funded Institution «The Joint Directorate of State Nature Reserves "Orenburg"» and «Shaitan Tau», Donetsk str., 2/2, Orenburg, Russia
russian969@yandex.ru, rbakirova@gmail.com
[3] Ufa State Petroleum Technological University, Kosmonavtov str., 1, Ufa, Russia
trioksan@mail.ru

Abstract. The comparison of seasonal dynamics of NDVI between years 2016 of normal-average rainfall and drought-affected 2010 was carried out in the Pre-Urals Steppe area (Orenburg State Nature Reserve, Russia), where a semi-free population of the Przewalski horse had been established. Landsat 7, 8 and Sentinel 2 satellite images were used to calculate NDVI. Vegetation productivity in 63 model-scientific plots were studied in between June 18 and 30, 2016 during the period of maximum development of the vegetation; NDVI for the plots were calculated too. The data were used to build a linear predictive model on the correlation between NDVI and vegetation productivity. Such modelling might prove effective in an estimation of pasture forage resources and the prediction of its changes in dry years. According to the model the extreme drought in 2010 resulted in a 60% decrease in vegetation productivity during the period of maximum development of the vegetation. After a severe drought the drop in winter forage resources may be much more drastic. Yet, a study of the depth and spatial distribution of snow cover is necessary for accurate predictions of a supply of pasture forage for the population of the Przewalski horses.

Keywords: NDVI · Steppe · Drought · Vegetation productivity · Pasture forage resources · Przewalski horse · *Equus ferus przewalskii*

© Springer Nature Switzerland AG 2019
I. Bychkov and V. Voronin (Eds.): *Information Technologies in the Research of Biodiversity*, SPEES, pp. 1–7, 2019.
https://doi.org/10.1007/978-3-030-11720-7_1

1 Introduction

Understanding species–habitat relationships is critical for wildlife management, providing information on habitat requirements, distribution, and potential land use impacts [1]. When establishing a semi-free population of rare hoofed animals in specially protected natural areas, one of the main challenges is a prediction of forage resources for grazing and their possible changes in years of extreme weather.

In 2015, the Russian Government gave a territory to Orenburg State Nature Reserve to found a new protected natural area called Pre-Urals Steppe (PUS). A semi-free population of the Przewalski horse *Equus ferus przewalskii* Polj., 1881 has been established in the area with the support of a UNDP/GEF project entitled "Improving the coverage and management efficiency of protected areas in the steppe biome of Russia". In 2016, vegetation mapping was done and productivity of grassland in the main types of the vegetation was analyzed by a harvest method. A GIS map of vegetation was created and productivity of grassland for grazing was calculated for 2016 which had favourable weather conditions [2].

The PUS area sometimes faces extreme droughts; the latest one happened in 2010. Thus, it becomes necessary to predict changes in forage resources for grazing in dry years. As field surveys were not carried out in this territory in dry years, a retrospective analysis of the impact of drought on vegetation may be used as the most suitable tool for such a prediction, on the basis of vegetation indices. Many researches addressed to evaluation of the impact of drought on ecosystem primary productivity, using various vegetation indices derived from remote sensing data [3]. It is difficult to decide what index for the evaluation of the impact of drought is the most reliable as ground-based surveys often cannot provide sufficient information to verify the drought indices obtained from a satellite [4]. Nonetheless, many studies show that the Normalized Difference Vegetation Index (NDVI) can be used as a surrogate for ecological productivity, and in fact reflects net primary productivity (NPP) of an ecosystem because it is highly associated with photosynthetically active radiation that drives photosynthesis [5–7].

2 Materials and Methods

Pre-Urals Steppe (PUS) is a fenced area measuring approximately 16×14 km and totalling 16538 ha. The centre of the area is located at $51° 10' 57.36''$N and $56° 10' 54.12''$E. Grasslands occupy more than 90% of the territory and represent 24 types of rich bunchgrass, psammophytic, petrophytic, halophytic steppes, and their anthropogenic derivatives [2].

Landsat-8 and Sentinel-2 satellite images were used to analyse the normal 2016; cloudless Landsat-5 images were used to analyse the dry 2010. NDVI was calculated for the total grassland area. The most of the images were cloudless; a few of them had 5–7% of cloud cover in the study area. When NDVI was calculated, pixels in cloud areas were not used.

In 2016, vegetation productivity in a series of 63 representative model-scientific plots was studied in between June 18 and July 1 during the period of maximum development of the vegetation (the end of blossoming and the beginning of the fruiting phenological phase of gramineous plants). Within each of the model-scientific plots vegetation was clipped in 3 areas of 1 m^2 in size each located 10–15 m from each other. Clipped vegetation was air-dried and weighted with accuracy to 0.01 g. The pasture forage resources of PUS totalled 434386.6 centners [2].

Correlation analysis on the relationship between productivity and NDVI of the vegetation in the model-scientific plots was carried out using three cloudless Sentinel-2 images taken on June 17, June 20, and July 7, 2016. Besides, correlation between productivity and the average of the three NDVI on the above-mentioned days. In doing this, the average NDVI for three images was calculated for each pixel located at coordinates of a model-scientific plot. Then, a regression technique was used to build a linear predictive model of vegetation productivity according to NDVI. The regression equations were used to calculate the pasture forage resources in PUS. For this, the average vegetation productivity and the average amount of pasture forage in an area totaling the size of a pixel were calculated; this value was then multiplied by the number of pixels for the total area of PUS. A linear model calculated from the average NDVI values from three satellite images of 2016, was used to estimate pasture forage resources in the same period of the maximum vegetation development in 2010. The calculation was made from the image of June 27, 2010.

To assess climate differences between 2010 and 2016 weather data obtained from the meteorological station No 35127 which is located in the city of Akbulak, Orenburg Region, 35 km south-west of PUS was used [8] (Table 1).

Table 1. Some climatic features of grass vegetation periods in 2010 and 2016 and its comparison with 50-year averages.

Year	Month						
	IV	V	VI	VII	VIII	IX	X
	Average temperature, °C						
2010	7.7	18.2	24.8	26.4	25.3	16.0	4.7
2016	10.0	16.0	19.7	22.5	26.0	13.7	n.a.*
50-year averages	7.5	15.7	20.7	22.6	20.8	14.1	5.3
	Monthly precipitation, mm						
2010	19.0	1.0	3.0	9.0	11.0	3.0	13.0
2016	25.3	48.8	13.5	23.6	2.3	78.6	n.a.*
50-year averages	26.0	29.0	38.0	33.0	24.0	27.0	30.0

* – not available

3   Results and Discussion

The dynamics of NDVI for grass vegetation during the drought season in 2010 and the season of normal-average rainfall in 2016 were analyzed (Fig. 1).

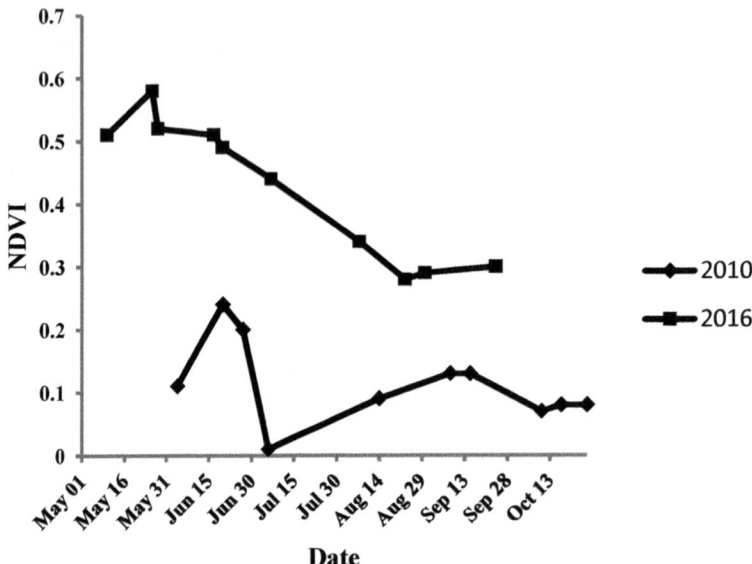

Fig. 1. The dynamics of NDVI for grass vegetation during growing seasons in 2010 and 2016.

In general, NDVI values during growing seasons were significantly lower in 2010 than in 2016. The differences were principally due to the amount of precipitation and to a lesser extent, to temperatures. During growing season in 2010, mean monthly temperatures were higher and precipitation were lower than 50-year averages (Table 1). Efficient precipitation did not appear in May and June, 2010 when grasses intensively grew. Mean monthly precipitation during growing season in 2016 was not different from 50-year averages; the values exceeded monthly precipitation of 2010 except of August. During the winter period between December 1, 2009 and March 30, 2010 the territory received 115.2 mm of precipitation. Between December 1, 2015 and March 30, 2016 the territory received 219.1 mm of precipitation which came mostly in the form of snow; this led to increase the soil moisture in the beginning of the seasonal vegetation development.

There was some decline in NDVI values during the period of maximum vegetation productivity in the last ten days of June, 2016. This appeared to be a result from the fact that the maximum photosynthetic activity of leaves of grasses normally coincides with the completion of formation of the leaves and it decreases by the end of blossoming. NDVI values continued decreasing in August and remained at the same level in September 2016.

In May and June, 2010 due to the drought the intensive vegetation development progressed in places with better moistening; in the beginning of July, green grasses partially survived only in ravines and along the beds of transient watercourses. In August and September, the NDVI values slightly increased due to rainfalls but remained almost three times less in comparison with the same period in 2016. Correlation coefficients between vegetation productivity in the model-scientific plots and the NDVI values for these plots were calculated (Table 2).

Table 2. Linear models for calculation of vegetation productivity in the pre-urals steppe area based on NDVI from the images taken on the days when the vegetation were clipped.

Date	Linear models	Correlation coefficient productivity (g/m², air dry weight) and NDVI
2016.06.17	Productivity (g/m) = 2.37231 + 564.203 *NDVI	0.75
2016.06.20	Productivity (g/m) = 8.89615 + 571.83 *NDVI	0.76
2016.07.07	Productivity (g/m) = 8.00107 + 653.782 *NDVI	0.77
Averages for three dates	Productivity (g/m) = −5.84523 + 616.156 *NDVI	0.77

In an earlier study in semi-arid areas, a high correlation was found between dry vegetation masses and Vegetation Indexes, i.e. the STI (0.66) [9] and iRVI (0.69) [10]. However, correlations between vegetation productivity and NDVI values were greater than or equal to 0.75 in the present study and exceeded the above-mentioned correlation.

These equations were used to calculate pasture forage resources in PUS in 2016 (Table 3).

Table 3. The comparison between the estimation of pasture forage resources in PUS carried out by NDVI and by the traditional harvest method.

Date	NDVI	Pasture forage resources (centners) calculated by NDVI	Differences (%) between pasture forage resources calculated by NDVI and by the harvest method
2016.06.17	0.513	482473	11.1
2016.06.20	0.492	479782	10.5
2016.07.07	0.436	485808	11.8
Average for three dates	0.480	479489	10.4

The NDVI values were declining during the period of harvesting of vegetation (Table 3). After harvesting was ended, on July 7, vegetation, and especially feather-grasses, turned yellow in some habitats where the phenological development progressed quicker (on sandy and stony soils). However, despite the decline in the average NDVI, pasture forage resources in dry weight did not decrease. The amount of forage resources estimated by regression equations using NDVI from 3 images surpassed the results obtained from the harvest method by 10.4–11.8% (479,489.0 centners vs. 434,386.6 centners). Thus, the use of NDVI permits accurate estimation of forage recourses in rather large pasture areas, albeit the estimation is somewhat overstated. In 2010, the estimated pasture forage resources were 1,903,978 centners, accounting for 39.7% of the resources during the same vegetation period in 2016.

To estimate pasture forage resources in the end of vegetation period, a linear model based on grass productivity during that period is necessary because the ratio between plant species with different loss of weight through drying changes.

NDVI values for the end of vegetation periods in 2010 and 2016 differed almost three times accordingly. While the proportion of green plants declines during severe drought, the significant amount of above-ground parts of motley grasses and even gramineous plants die and are dispersed by the wind by the end of vegetation period. It is therefore expected that pasture forage resources can decrease up to three times by the end of vegetation period during drought in comparison with a year with favourable weather conditions.

4   Conclusion

The calculation of grassland productivity by the linear models built on NDVI can be effective in estimating of pasture forage resources and predicting of their changes during droughts without mapping of plant communities. However, additional field studies by the traditional harvest methods in the main types of vegetation are necessary for precise estimation of pasture forage resources in the end of vegetation period. The projected decrease in grass productivity to three times in the end of vegetation period does not fully reflect the decrease in winter forage resources for the Przewalski horse population. This is because the plants growing on higher ground are the most available for the horses in snow winter, but it is the vegetation that is primarily affected by drought. Plants are better preserved in lowlands but they may be not available because of deep snow cover.

Thus, surveys of the depth and spatial distribution of snow cover in PUS is necessary for accurate predictions of the amount of pasture forage resources for the population of Przewalski horses in winter. Nevertheless, it is clear that a severe drought could reduce the pasture resources more than three-fold. Therefore, in the case of drought adequate supplies of piled hay must be stocked to guarantee Przewalski horse welfare and preservation of steppe ecosystem in PUS.

Acknowledgements. We greatly acknowledge Luibov G. Linerova, Vladimir Yu. Petrov, and Aleksei A. Kozyr for their kind technical and practical assistance.

References

1. Michaud, J.-S., Coops, N.C., Andrew, M.E., Wulder, M.A., Brown, G.S., Rickbeil, G.J.M.: Estimating moose (Alces alces) occurrence and abundance from remotely derived environmental indicators. Remote Sens. Environ. **152**, 190–201 (2014). https://doi.org/10.1016/j.rse.2014.06.005
2. Fedorov, N.I., Mikhailenko, O.I., Zharkikh, T.L., Bakirova, R.T.: Mapping of vegetation with the geoinformation system and determining of carrying capacity of the Pre-Urals Steppe area for a newly establishing population of the Przewalski Horse Equus ferus przewalskii at the Orenburg State Nature Reserve. IOP Conf. Ser.: Earth Environ. Sci. **107**, 012100 (2018). https://doi.org/10.1088/1755-1315/107/1/012100

3. Zhang, A., Jia, G.: Monitoring of meteorological drought in semi-arid regions using a multi-sensor microwave remote sensing data. Remote Sens. Environ. **13**, 12–23 (2013). https://doi.org/10.1016/j.rse.2013.02.023

4. Bayarjargal, Y., Karnieli, A., Bayasgalan, M., Khudulmur, S., Gandush, C., Tucker, C.J.: A comparative study of NOAA–AVHRR derived drought indices using change vector analysis. Remote Sens. Environ. **105**, 9–22 (2006). https://doi.org/10.1016/j.rse.2006.06.003

5. Numata, I., Roberts, D.A., Sawada, Y., Chadwick, O.A., Schimel, J.P., Soares, J.V.: Regional characterization of pasture changes through time and space in Rondonia. Brazil. Earth Interact. **11**, 1–25 (2007). https://doi.org/10.1175/EI232.1

6. Numata, I., Soares, J.V., Roberts, D.A., Leonidas, F.C., Chadwick, O.A., Batista, G.T.: Relationships among soil fertility dynamics and remotely sensed measures across pasture chronosequences in Rondonia. Brazil. Remote Sens. Environ. **87**, 446–455 (2003). https://doi.org/10.1016/j.rse.2002.07.001

7. Li, Z., Huffman, T., McConkey, B., Townley-Smith, L.: Monitoring and modeling spatial and temporal patterns of grassland dynamics using time-series MODIS NDVI with climate and stocking data. Remote Sens. Environ. **138**, 232–244 (2013). https://doi.org/10.1016/j.rse.2013.07.020

8. Specialized sets for climate studies, http://aisori-m.meteo.ru/waisori. Last accessed 13 Aug 2018

9. Jacques, D.C., Kergoat, L., Hiernaux, P., Mougin, E., Defourny, P.: Monitoring dry vegetation masses in semi-arid areas with MODIS SWIR bands. Remote Sens. Environ. **153**, 40–49 (2014). https://doi.org/10.1016/j.rse.2014.07.027

10. Verbesselt, J., Somers, B., van Aardt, J., Jonckheere, I., Coppin, P.: Monitoring herbaceous biomass and water content with SPOT VEGETATION time–series to improve fire risk assessment in savanna ecosystems. Remote Sens. Environ. **101**, 399–414 (2006). https://doi.org/10.1016/j.rse.2006.01.005

Regional Information and Analytical System on Insect Biodiversity of the Baikal Region

Igor A. Antonov(✉) (iD)

Siberian Institute of Plant Physiology and Biochemistry of Siberian Branch of
Russian Academy of Sciences, Lermontova Str. 132, 664033 Irkutsk, Russia
patologi@sifibr.irk.ru

Abstract. Regional information and analytical system for assessing and monitoring of insect biodiversity in the Baikal region has been created. This system is consisting of three blocks: (a) relational databases, (b) the geoinformation system, (c) the software environment R. Relational databases are designed to quickly find the necessary information and to automate the collection and structuring of data. Besides, these databases were created to analyze the requested information and to create an information basis for geoinformation system. In the structural and semantic respect, the databases consist of several parts: taxonomic, ecological, geographic. The geoinformation system was created to study the spatial distribution of insects. It covers the territory of three subjects of the Russian Federation (Irkutsk oblast, Republic of Buryatia and Zabaykalsky krai). All cartographic information in the geoinformation system is organized in the form of vector and raster layers: base (topological) layers, including rivers, roads, buildings, etc.; thematic layers (for example, entomological and landscape). Raster layers (cosmic images and digitized paper maps) were used when creating base vector layers. R is one of the most popular platforms for statistical computing and graphics, because it is free, open-source software, with versions for Windows, Mac OS X, and Linux operating systems. At the present time, the features of spatial distribution of ants and buprestids have been explored by means of this information and analytical system.

Keywords: Relational databases · Geoinformation system · Software environment R

1 Introduction

The problem of biodiversity conservation is topical for modern ecology, because the scale of the impact of anthropogenic press on natural ecosystems is expanding at an accelerated pace. Besides, it is very important to study the spatial distribution of species. Therefore, the use of geoinformation systems (GIS) in the biodiversity investigation becomes more extensive. GIS allows more efficient use of extensive data arrays, accelerate and objectify the process of creating maps and expand the scope of their practical application [1].

The Baikal region has complex relief, heterogeneity of climate, large diversity of soil types and plants. This complex set of parameters, together with the large area of

© Springer Nature Switzerland AG 2019
I. Bychkov and V. Voronin (Eds.): *Information Technologies
in the Research of Biodiversity*, SPEES, pp. 8–13, 2019.
https://doi.org/10.1007/978-3-030-11720-7_2

this region, creates a large variety of habitats and has a distinct influence on the biodiversity of insects which play a huge role in the terrestrial ecosystems of the Baikal region [2]. Therefore, the information and analytical system for assessing and monitoring of insect biodiversity in the Baikal region has been created.

2 The Structure of Regional Information and Analytical System

This system is consisting of three blocks: (a) relational databases, (b) the geoinformation system, (c) the software environment R (see Fig. 1).

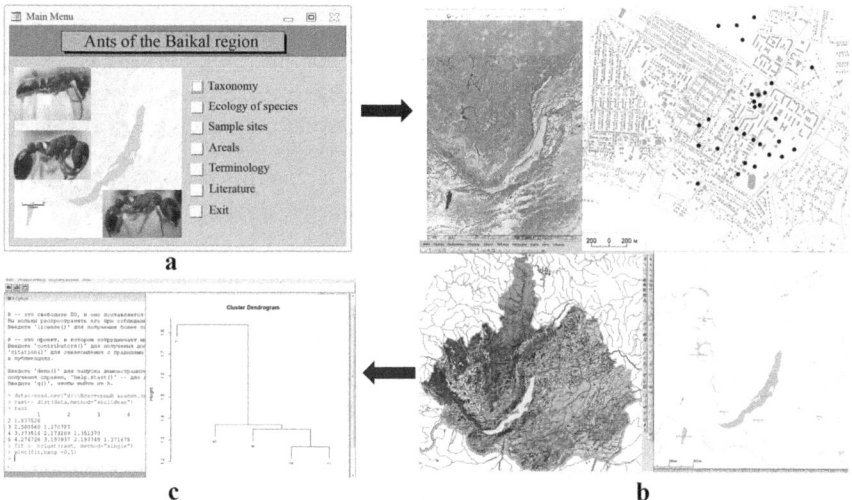

Fig. 1. The structure of regional information and analytical system (explanations in the text).

Relational databases are designed to quickly find the necessary information and to automate the collection and structuring of data. Besides, these databases were created to analyze the requested information and to create an information basis for geoinformation system. In the structural and semantic respect, the databases consist of several parts: taxonomic, ecological, geographic. The taxonomic part includes information on the species composition of entomofauna of the Baikal region, taking into account recent changes in the taxonomy. The ecological part includes a description of the ecology of each species, as well as a description of the type of species' habitat. The geographic part contains data on the place of collection (district, subject of the Russian Federation, description of location), geographical coordinates and arealogical characteristics of the species.

The geoinformation system was created to study the spatial distribution of insects. It covers the territory of three subjects of the Russian Federation (Irkutsk oblast, Republic of Buryatia and Zabaykalsky krai). All cartographic information in the geoinformation system is organized in the form of vector and raster layers: base (topological) layers, including rivers, roads, buildings, etc.; thematic layers (for example, entomological and landscape). As a result, GIS works with two essentially different types of data (vector and raster). The vector model is especially useful for describing discrete objects and is less suitable for describing continuously changing properties [1]. The raster model is optimal for working with continuous properties [1]. Raster layers (cosmic images and digitized paper maps) were used when creating base vector layers. Moreover, due to the development of cartographic services on the Internet (access to high-resolution cosmic images, about 1 m per pixel), vector layers with a wide scale range were created, which makes it possible to create both small-scale and large-scale thematic maps. This allows studying a biodiversity at different levels.

R is one of the most popular platforms for statistical computing and graphics, because it is free, open-source software, with versions for Windows, Mac OS X, and Linux operating systems. R is a case-sensitive, interpreted language [3]. Besides, R comes with extensive capabilities right out of the box, but some of its most exciting features are available as optional modules (over 2.500 user-contributed modules) that are downloaded from the repository (http://cran.r-project.org/web/packages) [3].

3   The Results of Using the Information Analytical System

At the present time, the features of spatial distribution of ants and buprestids have been explored by means of this information and analytical system [4, 5]. Ants run much of the terrestrial world as the premier soil turners, channelers of energy and dominatrices of the insect fauna [6]. The buprestids form the eighth largest family (nearly 14,900 valid species) of Coleoptera with many extant species common and abundant in their respective habitats [7]. Besides that, buprestids are important pest of forest. Buprestids were registered on burned-out forest areas, they damages coniferous and leaf-bearing trees, and these insects are not only the technical pests, but also they are capable to do physiological damage to trees [5].

In the research of the intraspecific diversity of ant *Myrmica angulinodis*, the map of the study region and locations of *M. angulinodis* population groups was created using GIS (see Fig. 2). The base (topological) vector layers and the vector layer of sample sites of ants were used in this map. The effect of geographic origin on morphometric characteristics of ants has been studies by means of one-way ANOVA using 120 worker specimens of *M. angulinodis* from four population groups inhabiting different parts of the Baikal region, and similarity between the population groups has been evaluated using cluster analysis (see Fig. 3). All statistical processing was performed in the software environment R. The "MASS", "ggdendro" and "vegan" R-packages were used. Thus, the results of analysis show that worker specimens in populations from areas with lower air temperatures have significantly greater body size and that ants with an elongated head and more strongly curved frontal carinae dominate in the south of Irkutsk oblast, while ants with a wider frons prevail in the south of the Republic of Buryatia [4].

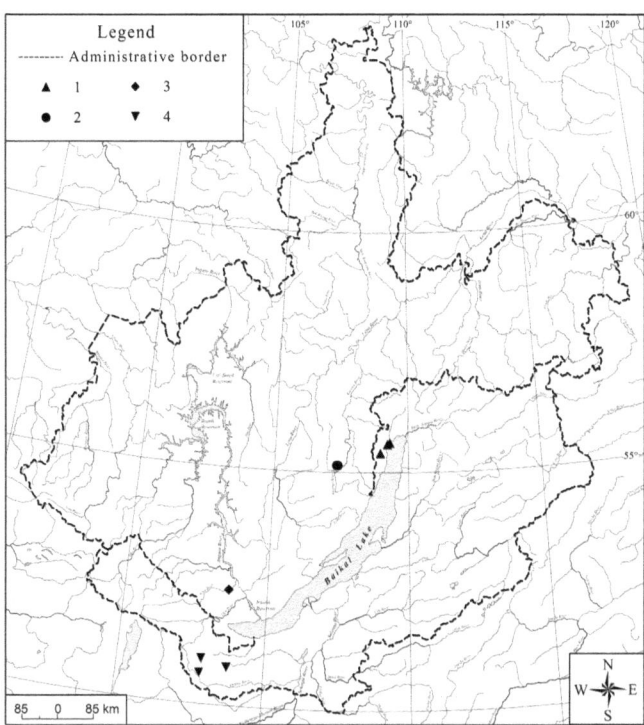

Fig. 2. Map of the study region and locations of *Myrmica angulinodis* population groups: 1 – Northern Buryatia; 2 – Central Irkutsk oblast; 3 – Southern Irkutsk oblast; 4 – Southern Buryatia.

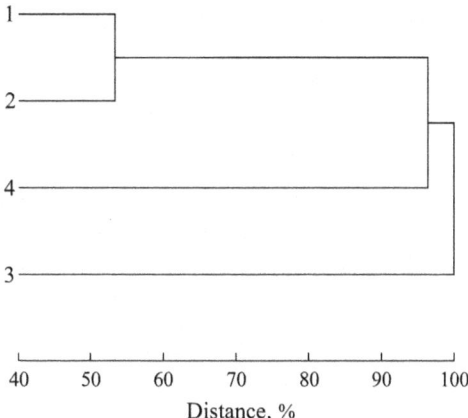

Fig. 3. Dendrogram of similarity between *M. angulinodis* population groups (1–4, see Fig. 1) based on complete linkage (furthest-neighbor) clustering of Euclidean distances and scaled with respect to the maximum distance.

In the study of buprestids, the following results were obtained using the information and analytical system [5]:

1. The map of specimen collection sites of 11 buprestids species of coniferous trees in territories of the Baikal region and Northern Priamur'e was created for the first time (see Fig. 4).

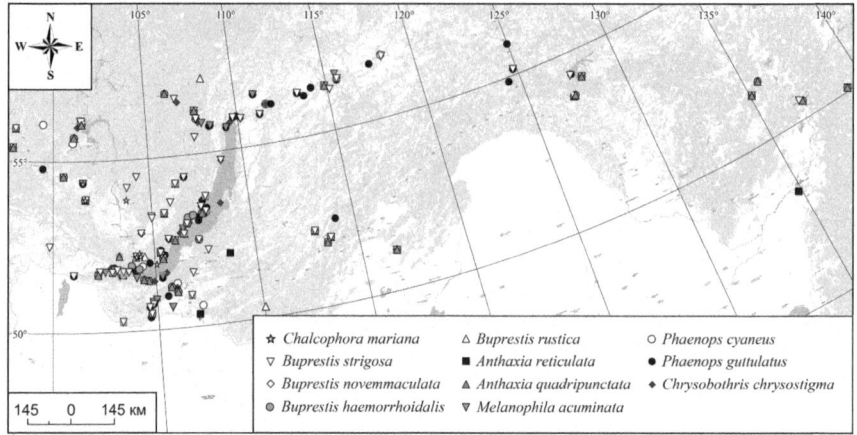

Fig. 4. Map of specimen collection sites of 11 species of coniferous trees.

2. Two species (*Anthaxia psittacina* and *Chrysobothris amurensis*) were recorded in the fauna of Baikal region for the first time.
3. The buprestids with wide ranges (holarctic and transpalaearctic) prevail and their part is 43.5%.
4. Five species (*Anthaxia quadripunctata, Buprestis strigosa, B. haemorrhoidalis, Chrysobothris chrysostigma* and *Phaenops guttulatus*) have high frequency of occurrence.
5. Faunas of buprestids of the Republic of Buryatia and Zabaykalsky Krai were most similar (coefficient of similarity Chekanovsky-Serensen is equal to 0.8).

4   Conclusion

The role of information and analytical systems is constantly growing, because only with their help can promptly consider the latest changes in the environment (the influence of the anthropogenic factor) and in the taxonomy of plants and animals. In our researches, the information and analytical system showed high efficiency in the study of the insect biodiversity in the territory of the Baikal region.

Acknowledgements. The author is grateful to Prof. A. G. Radchenko (Schmalhausen Institute of Zoology, National Academy of Sciences of Ukraine, Kiev) for his help in identification of worker specimens from different *M. angulinodis* colonies, to Dr. A. A. Sorokovoy (V. B. Sochava Institute of Geography SB RAS, Irkutsk) for his assistance in the mastering of GIS technology, and to Dr. Yu. S. Bukin (Limnological Institute, Siberian Branch, Russian Academy of Sciences, Irkutsk) for his assistance in getting familiar with the software environment R.

This study was conducted within the project no. AAAA-A17-117011810101-8.

References

1. Tikunov, V.S.: Modeling in Cartography. MGU, Moscow (1997). (in Russian)
2. Antonov, I.A., Pleshanov, A.S.: Landscape ecological complexes of ants in Baikalian Siberia. Sibirskii Ekologicheskii Zhurnal **15**(1), 53–57 (2008). (in Russian)
3. Kabacoff, R.: R in Action: Data Analysis and Graphics with R, 1st edn. Manning Publications Co., New York (2011)
4. Antonov, I.A.: Interpopulation variation in morphometric characteristics of the ant *Myrmica angulinodis* Ruzs. (Hymenoptera: Formicidae) in the Baikal region. Russ. J. Ecol. **48**(4), 358–363 (2017). https://doi.org/10.1134/s106741361704004x
5. Agafonova, T.A., Silaev, A.S., Antonov, I.A.: The Analysis of Spatial Distribution of Buprestids (Coleoptera: Buprestidae) in the Baikal Siberia and Northern Priamur'e. Bull. Irkutsk State Univ.: Ser. «Biol. Ecol.» **20**, 37–46 (2017). (in Russian)
6. Hölldobler, B., Wilson, E.O.: The Ants. Harvard University Press, Cambridge (1990)
7. Xiaoxiong, P., Huali, C., Dong, R., Chungkun, S.: The first fossil buprestids from the Middle Jurassic Jiulongshan formation of China (Coleoptera: Buprestidae). Zootaxa **2745**, 53–62 (2011)

Use of Remote Sensing Data and GIS Technologies for Monitoring Stocks of Medicinal Plants: Problems and Prospects

N. B. Fadeev[1]([⊠]) [ID], T. N. Skrypitsyna[2] [ID], V. M. Kurkov[2] [ID], and N. I. Sidelnikov[1] [ID]

[1] All-Russian Research Institute of Medicinal and Aromatic Plants, 7, Grina Street, 117216 Moscow, Russia
nfadeev@mail.ru
[2] Moscow State University of Geodesy and Cartography, MIIGAiK, 4, Gorokhovsky Pereulok, 105064 Moscow, Russia
mola_mola@rambler.ru, vkurkov@inbox.ru

Abstract. Preservation of the biodiversity of medicinal plants makes it possible to create a basis for the development and creation of new medicines through scientifically based nature management. Remote sensing techniques provide solutions for obtaining, analyzing and managing data at different scales: high-resolution satellite imagery and multi-spectral channels can provide information covering large areas, while aerial photography with unmanned aerial vehicles (UAV) allows you to collect comprehensive biometric information from key areas. Studies carried out on the territory of the Tula region in 2012–2016, showed the advisability of using the UAV for monitoring the reserves of wild medicinal plants. The planned accuracy of the created orthophoto, their photographic quality allowed confidently interprets the plant communities during flowering. Plant resources were estimated with an accuracy of not less than 10%. The result of the joint work was a map and database of stocks of medicinal plants for the study site. The use of multispectral aerial and space imagery, in addition to aerial photography data in the visible range, scales the research of medicinal plants resources on significant areas, which should accelerate the work on large areas. Nevertheless, one of the important problems remains the automation of interpretation of wild-growing herb medicinal plants, connected with the peculiarity of the growth of these communities. The solution of the problem is the application of complex analysis using digital terrain models.

Keywords: Biodiversity · Remote sensing · Plant interpretation

1 Introduction

Preservation of the biodiversity of medicinal plants makes it possible to create a basis for the development and creation of new medicines through scientifically based nature management. One of the modern tasks of botanical resource research is the interpretation of plant communities, monitoring their condition and spread. An example of creating a database is the GIS "Dikorosy" application of the geoportal of the Tomsk

© Springer Nature Switzerland AG 2019
I. Bychkov and V. Voronin (Eds.): *Information Technologies in the Research of Biodiversity,* SPEES, pp. 14–21, 2019.
https://doi.org/10.1007/978-3-030-11720-7_3

region, which was created in the TSU, which allows forecasting the yields of cranberries, lingonberries, blueberries on the basis of data on yields of wild plant resources, as well as meteorological data by seasons of the year, pine nuts [1].

Biomass is a key indicator of the ecosystem and, in particular, of medicinal plant communities. An accurate assessment using remote sensing is important for understanding the response of meadow communities to climate change [2] and anthropogenic impact [3].

Remote sensing techniques provide solutions for obtaining, analyzing and managing data at different scales: high resolution satellite imagery and different spectral channels can provide information covering large areas [4], while aerial photography from unmanned aircraft (UAV) allows the assembling a different of biometric information from key areas [5, 6].

2   Characteristics of the Object

The research was carried out on the territory of the Zaoksky test site of the Moscow State University of Geodesy and Cartography in the Tula region.

In botanico-geographical terms, the landfill site is a system of elevations and depressions in the flood plain of the Sknigi River. Within its boundaries are landscapes typical of many areas of Central Russia. On the territory of the geopoligon there are various forms of relief of various genesis, hydrographic objects (rivers, streams, ponds), forest tracts, agricultural lands.

The following medicinal plants were interpreted: fild horstail (*Equisetum arvense* L.), yellow bedstraw (*Galium verum* L.), meadowsweet (*Filipendula ulmaria* (L.) Maxim.), echinops (*Echinops sphaerocephalus* L.), nettle (*Urtica dioica* L.), fireweed (*Chamaenerion angustifolium* (L.) Scop.), cats tail (*Typha angustifolia* L.), che reed grass (*Calamagrostis epigeios* (L.) Roth), angelica (*Archangelica officinalis* Hoffm.).

3   Methods and Technologies

The article considers the experience of using various methods of remote sensing for the purpose of interpretation and evaluation of wild medicinal plants. The processing was carried out both from high-resolution satellite images obtained from the satellite WORLDVIEW-2 (DigitalGlobe, Inc.) using aerial photography.

For the purposes of aerial photography of communities, various airborne monitoring systems, aerial photographs with the participation of photogrammetric methods to be processed have been carried out for a number of years (2012–2017). The best-quality shooting of the 2014 from the Orlan-10 unmanned vehicle was used to determine areas and create interpretation standards. The photos were taken by the PhaseOne iXU camera from a height of 250 m, which makes it possible to obtain aerial photographs with a resolution on the ground up to 2.5 cm. These data are fundamental for monitoring.

To perform the work we used a complex of Digital Photogrammetric Workstation PHOTOMOD, GIS MapInfo Professional 12, a free geographic information system with open source QGIS. The result of the joint work was a map and a database of stocks of medicinal plants for the study site.

3.1   Stereo Interpretation as a Method of Processing Data to Create Standards for the Key Area

As you know, there is already experience in applying stereo decoding and forest mapping in aerial photographs [7, 8]. With the communities of wild grassy cases, no work was done, because of their small size and high mosaic of the image.

Nevertheless, aerial photography from the UAV made it possible to recognize a number of tall grass plants during their flowering period. As a result, photogrammetric processing, stereo interpretation, visual and metric analysis of the areas of medicinal plants.

By stereo images it was possible to determine the biometric characteristics, such as the height of the plant and the number of individuals per square meter, which led to a significant reduction in field work. The stock of raw materials was determined with an accuracy of not less than 10%. Based on the results of the interpretation, a digital map of the areas of medicinal plants was created and a database with metric indicators of bioproductivity [5]. This method is relevant in cases of monitoring stocks of medicinal plants in hard-to-reach areas (mountain massifs, water objects, wetlands).

3.2   Monitoring Stocks of Medicinal Plants

The basis for visual analysis was the photoplanes and stereo models obtained as a result of the series of surveys 2012–2016. Are compared areas of communities, the density of growth. As a result of the conducted studies, experimental data on aerial photography of plant communities with the use of artificial intelligence were obtained, digital spatial data on stocks of medicinal plants taking into account the relief (dynamics for 5 years of observations were estimated), which makes it possible to carry out an economic assessment of plant resources in a particular area. The assessment was carried out taking into account the weather conditions that were during the filming.

The cartographic basis for the studied territory was created, nine types of medicinal plants were noted (2012 - 91 areas, 2014-115 areas, in 2016 - 76 areas, the total area of medicinal plants in 2012 was 11,713 m^2, 2014 g .- 30,950 m^2, in 2016 – 7579 m^2).

On the example of some medicinal plants, let us consider the possibility of decoding and monitoring of stocks using the UAVs.

Meadowsweet (*Filipendula ulmaria*) (Fig. 1): in 2016 we observe a smaller number of inflorescences, perhaps the plant is not in the full flowering phase. The quality of the initial photos of 2016 is worse than in 2014 (smear, the resolution on the terrain is worse), therefore, the measurement error and the merging of images in social colors increase at the same time in 2016, when the density of generative forces (inflorescences) obtained by visual cameral interpretation is two times less than in 2014.

According to the picture, we see that the range of growth of the meadowsweet has generally remained within its borders. As this medicinal plant grows in moist, low-lying places, we can conclude that the humidity in this place has not changed in two

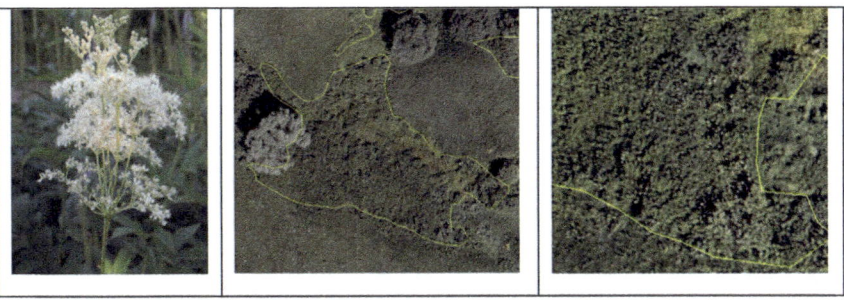

Fig. 1. Meadowsweet (*Filipendula ulmaria*) - pattern areas, stereo image interpretation.

years. Meadowsweet forms stable communities, occupying the territory for more than a hundred years, due to the fact that it is a rhizome with a high community density and high projective coverage.

It is worth noting that the reference survey of 2014 illustrates an earlier period of vegetation and flowering, and the shooting of 2012 was performed in a later period of vegetation - fruiting (August), so it was not possible to calculate the number of inflorescences per m^2.

Below (Figs. 2 and 3) the monitoring data of medicinal plant communities is presented.

Vectorization and monitoring of the 2012 survey: The pink outline shows the boundaries of the communities identified in the survey of 2014 and serve as a reference, and the yellow outline shows the boundaries of the communities identified in the 2012 survey.

Fig. 2. Interpretation of the Meadowsweet (*Filipendula ulmaria*) (survey of 2014).

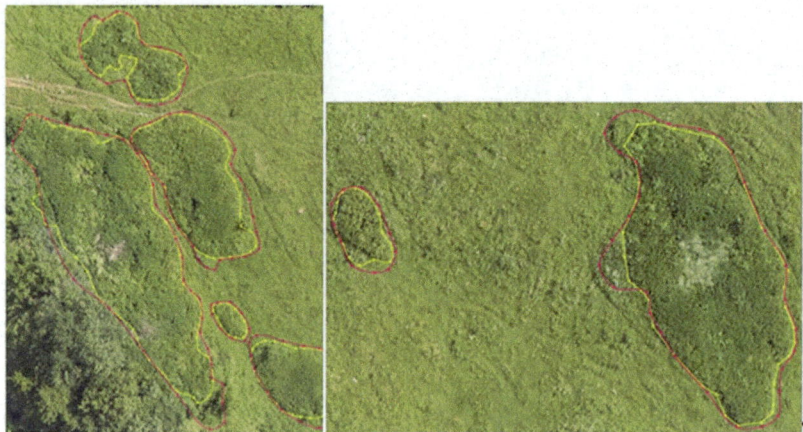

Fig. 3. Interpretation of the Nettle (Urtica dioica) (survey of 2014).

Nettle: it remained within the boundaries of its areas, which is explained by the fact that cattle grazing is actively carried out on this territory, as well as trampling by man. These factors do not allow it to spread and expand its local area.

The suggested technique allows creating a geobotanical map, where as a carto-graphic basis accurate, actual orthophoto are used, which, in turn, allow for desk interpretation. The use of photographic plans in the GIS environment makes it possible to create databases that contain quantitative and qualitative characteristics of plants. The created methodology can become part of the technology of natural resources assessment and forecasting of the state of the lithosphere and the biosphere for sus-tainable nature management.

However, visual interpretation is a fairly time-consuming process when it is nec-essary to analyze the communities of wild medicinal plants in large areas.

3.3 The Experience of Applying Multispectral Images Obtained with the UAV for Automatic Interpretation

The purpose of these studies was to test and evaluate the possibility of automatic interpretation of medicinal plants based on multispectral data obtained from the UAV.

The tasks assigned to achieve the goal: obtaining a synthesized multispectral orthophoto from RGB and NIR in a cartographic coordinate system; definition of reference areas; automatic image classification; evaluation of classification accuracy; creation of a vector map from the received data.

At the initial stage of the research, the feasibility of applying UAV data to create a multispectral image for further thematic processing was conducted. The results of the work showed that the creation of a multichannel image from the data of different surveys is possible, but because of the cut-off situation it is not possible to create a homogeneous illuminance orthophoto. For deeper processing, it is necessary to cal-culate the reflectivity of the surface, which is possible if information is available on the sensitivity of each channel of the camera, the capacity of the lens, the illumination of the underlying surface, and the exact time taken for photographing.

In connection with this, the next step in the research was the use of data obtained from ultra-high resolution spacecraft, in this case "WORLDVIEW-2".

3.4 The Use of Multispectral Images Obtained from Spacecraft for Automatic Interpretation

The cameras of many spacecraft are calibrated and their characteristics are known, while the spacecraft can shoot a large area in a few seconds for a few seconds, and the stereopair of the images is obtained within a few minutes, that is, under the same illumination conditions. In addition, many data providers produce radiometric correction, including removing the influence of the atmosphere. Managed classification was carried out in the program complex "ENVI".

The purpose of the classification was to establish the area occupied by medicinal plants in the area of interest. Medicinal plants subject to classification: fild horstail (*Equisetum arvense* L.), yellow bedstraw (*Galium verum* L.), meadowsweet (*Filipendula ulmaria* (L.) Maxim.), echinops (*Echinops sphaerocephalus* L.), nettle (*Urtica dioica* L.), fireweed (*Chamaenerion angustifolium* (L.) Scop.), cats tail (*Typha angustifolia* L.), che reed grass (*Calamagrostis epigeios* (L.) Roth), angelica (*Archangelica officinalis* Hoffm.).

Processing was carried out in several stages:

1. Creation of reference areas from vector data;
2. Classification of orthorectified images;
3. Evaluation of classification accuracy;
4. Converting the resulting raster into vector data;
5. Calculation of the area covered.

Areas of growth were presented in the form of vector data obtained by interpretation in a stereo mode by aerial survey. From the data obtained, standards were created on a classified multi-channel orthophoto.

The classification was carried out using the algorithm "Neural networks". The results of the classification are shown in Fig. 4.

The data obtained after the classification was converted from a raster format to a vector file.

The result is presented in the form of vector data and an embedded multispectral orthophoto. The results of the calculation of the areas of growth of different plant communities are presented in Table 1.

The minimum area of a homogeneous vegetation community, which we managed to classify, was 4 m^2, and the total area of classified medicinal plant communities was 458 ha.

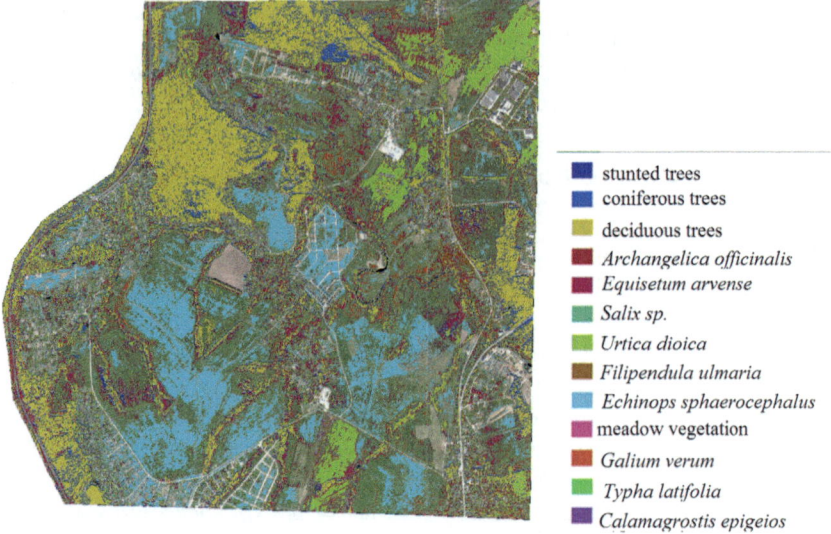

Fig. 4. The results of the classification by the method of "NeuroNet".

Table 1. Areas of growth of various plant communities within the studied area.

Class name	Area, hectares
Stunted trees	53.59
Coniferous trees	123.31
Seciduous trees	413.88
Meadow vegetation	316.14
Archangelica officinalis	6.14
Equisetum arvense	196.93
Salix sp.	0.28
Urtica dioica	174.92
Filipendula ulmaria	50.83
Echinops sphaerocephalus	2.68
Galium verum	58.93
Typha latifolia	6.65
Calamagrostis epigeios	19.79

4   Conclusions

The use of multispectral aerial and space imagery, in addition to aerial photography data in the visible range, scales the research of medicinal plants resources on significant areas, which should accelerate the work on large areas. Nevertheless, one of the important problems remains the automation of interpretation of wild-growing herb medicinal plants, connected with the peculiarity of the growth of these communities. The solution of the problem is the application of complex analysis using digital terrain models.

References

1. Khromykh, V.V.: The geo-portal of the Tomsk region on the basis of GIS for monitoring and forecasting the state of natural resources. In: Modern Problems of Geography and Geology: Dedicated to the Centenary of the Opening of the Natural Science Branch in Tomsk State University. pp. 112–117 (2017)
2. Huifang, Z., Yi, S., Li, C., Yu, Q., Jianjun, C., Yan, Q., Jiaxing, D., Shuhua, Y., Yingli, W.: Estimation of grassland canopy height and aboveground biomass at the quadrat scale using unmanned aerial vehicle. Remote Sens. **10**(851), 1–19 (2018)
3. Cornejo-Denman, L., Romo-Leon, J., Castellanos, A., et al.: Assessing riparian vegetation condition and function in disturbed sites of the arid Northwestern Mexico. Land **7**(1), 13 (2018). https://doi.org/10.3390/land7010013
4. Piyasinghe, I.P.K., Gunatilake, J., Madawala, H.M.S.P.: Mapping the distribution of invasive shrub Austroeupatorium inulifolium (Kunth) R. M. King & H. Rob: A case study from Sri Lanka. Ceylon J. Sci. **47**(1), 95–102 (2018). http://doi.org/10.4038/cjs.v47i1.7492
5. Fadeev, N., Skrypitsyna, T., Kurkov, V.: Modern geoinformation technologies in investigation of resources medicinal plants. Questions Biol. Med. Pharm. Chem. **6**, 68–73 (2016)
6. Fraser, R., van der Sluijs, J., Hall, R.: Calibrating satellite-based indices of burn severity from UAV-derived metrics of a burned boreal forest in NWT, Canada. Remote Sens. **9**, 279 (2017). https://doi.org/10.3390/rs9030279
7. Balenović, I., Seletković, A., Pernar, R., Jazbec, A.: Estimation of the mean tree height of forest stands by photogrammetric measurement using digital aerial images of high spatial resolution. Ann. For. Res. (2015). https://doi.org/10.15287/afr.(2015).300
8. Liu, H., Changshan, W.: Incorporating crown shape information for identifying ash tree species. Photogram. Eng. Remote Sens. **84**(8), 495–503 (2018). https://doi.org/10.14358/PERS.84.8.495(2018)

Digital Herbarium Collections of the Central Siberian Botanical Garden SB RAS, Novosibirsk, Russia

Nataliya Kovtonyuk$^{(\boxtimes)}$, Irina Han, and Evgeniya Gatilova

Central Siberian Botanical Garden SB RAS, Zolotodolinskaya str., 101, 630090
Novosibirsk, Russian Federation
nkovtonyuk@csbg.nsc.ru

Abstract. The first herbarium at the Central Siberian Botanical Garden (CSBG SB RAS) was organized in 1946. Now there are two herbarium collections in CSBG with their own acronyms and registrations in the Index Herbariorum (NSK and NS). Collection contains about 800,000 herbarium specimens of higher vascular plants, mosses, lichens and fungi sampled in Siberia, Russian Far East, Europe, Asia and America. Digitization of higher vascular plants at 600 dpi was initiated in 2014 by special scanner Herbscan starting with type specimens. Images and metadata of 889 type specimens are currently available on the Virtual Herbaria web site at the Vienna University, Austria and in JSTOR. In 2017 a new research group "USU-Herbarium" was organized in CSBG for digitization and management of herbarium collections. Our aim is to provide online access to CSBG SB RAS herbarium collections as a worldwide resource for biodiversity studies. We are digitizing NSK and NS high vascular plants collections by two scanners ObjectScan 1600 and Herbscan. Currently about 13,000 herbarium specimens were digitized at 600 dpi. Images and metadata are stored in CSBG Database generated by ScanWizard Botany and MiVapp Botany software (Microtek, Taiwan). The largest numbers of samples scanned are from Primulaceae (4984), Cystopteridaceae (891), Orchidaceae (750), Boraginaceae (664), Poaceae (632), Asteraceae (550), Athyriaceae (436), Fabaceae (418), Amaryllidaceae (381). The database is structured in a way that a user can access a high resolution image and following key information: specimen ID (= barcode), family name, scientific name, collector name and collection date, country or administrative region.

Keywords: Digitization · Herbarium database · Biodiversity

1 Introduction

"Species Plantarum" by Linnaeus, 1753, contained 5,940 species of plants, including all known species then known globally [1]. Since its publication 265 years ago, the exploration of plant diversity across the planet has led to approximately 374,000 known, described and accepted plant species [2]. The number of plant species known since then has increased more than 60 times. Herbarium collections are the most important source of scientific information about the distribution of objects in the past

© Springer Nature Switzerland AG 2019
I. Bychkov and V. Voronin (Eds.): *Information Technologies in the Research of Biodiversity*, SPEES, pp. 22–27, 2019.
https://doi.org/10.1007/978-3-030-11720-7_4

and the present, which allows simulation the dynamics of objects in the future. Only the herbarium sample reliably confirms the presence of the plant organism in a specific point of space in a certain period of time. Herbarium collections and the data they hold are valuable for more traditional studies of taxonomy and systematic, but also for ecology, bioengineering, conservation, food security, and the human social and cultural elements of scientific collection [3]. The value and universality of herbarium specimens are recognized in most countries, where national and large regional herbariums are actively developing and improving.

The creation of Virtual herbarium became a modern trend, stage of the inventory and modernization of herbarium collections of the leading Botanical institutions in the world [4–7]. Our aim is to provide online access to CSBG SB RAS herbarium collections as a worldwide resource for biodiversity study. Writing the long-awaited "Flora of Russia" would be greatly accelerated in the presence of network centers of Virtual herbarium collections of Botanical institutions in Russia.

The first herbarium at the Central Siberian Botanical Garden (CSBG SB RAS), Novosibirsk, Russia was organized in 1946. Now there are two herbarium collections in CSBG with their own acronyms and registrations in the Index Herbariorum (NSK and NS). Collection contains about 800,000 herbarium specimens of higher vascular plants, mosses, lichens and fungi sampled in Siberia, Russian Far East, Europe, Asia and America [8].

2 Materials and Methods

Digitization of higher vascular plants of NSK and NS collections was initiated by customized HerbScan unit, which consists of a flatbed scanner (Epson Expression model 10000XL), modified for inverted use, supported by Andrew Mellon Foundation in 2014–2016. Images and metadata of 889 NSK and NS type specimens are currently available on the Virtual Herbaria web site at the Vienna University, Austria (http://herbarium.univie.ac.at/database) and some of them in JSTOR (https://plants.jstor.org). Special attention was paid to providing on-line high resolution (600 dpi) images and metadata for all type specimens [5, 9, 10].

In 2017 a new research group "USU-Herbarium" was organized in CSBG SB RAS for digitization and management of herbarium collections (registered as USU_440537). We initiated digitization of NSK and NS higher vascular plants collections by two scanners ObjectScan 1600. Specimens were scanned using international standards: at a resolution of 600 dpi, the barcode for each specimen, 24-color scale and scale bar (see Fig. 1). Images and metadata are stored in CSBG SB RAS Database (http://84.237.85.99:8081), generated by ScanWizard Botany (reg. number I41-018966) and MiVapp Botany (reg. number I41-018969) software (Microtek, Taiwan). Specimen label information have recognized and automatically saved titled by herbarium code and specimen serial number in XML format through ScanWizard-Botany. MiVapp-Botany is both a web-server system and specimen image authentication database, aiming for being an efficient and integrated multi-functional platform. After hierarchical login-based image quality and metadata profile validation by experts, MiVapp-Botany can quickly update the system and make verified specimen access for users [11].

3   Results

Currently about 13,000 herbarium specimens from NS (4549) and NSK (8566) were digitized at 600 dpi. The database is structured in a way (see Fig. 1), that a user can access a specimen file that contains high resolution image and following key information: specimen ID (=barcode), family name, scientific name (genus, species, author of taxon), collector name and collection date, country or administrative region.

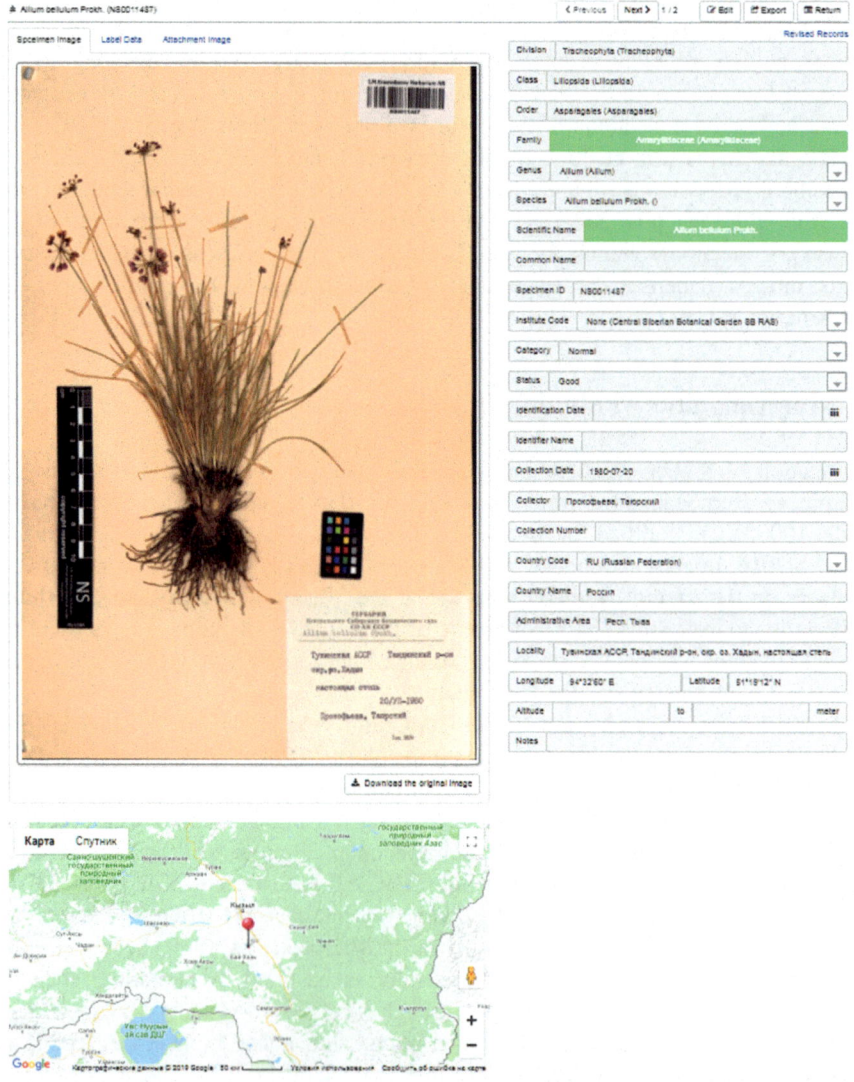

Fig. 1. CSBG SB RAS database with scanning herbarium specimen, map and metadata of Allium bellulum (NS0011508).

Alternatively the request may be done by using key words of habitat characteristics, for example, "pine forest", "meadows", etc. Images are available for download. In our internal database each image is supplied by the following information: barcode, type status, genus name, species name, author name, subsp/var name, family name, collectors name, field number of herbarium specimen, date (yyyy-mm-dd), country, admin region, latitude degrees, latitude minutes, longitude degrees, longitude minutes, label text, identifier name, annotation, Catalogue of Life link and accepted name in CoL.

Data from herbarium labels are entered in 26 fields in the Calc tables of the LibreOffice software package. Posting information from herbarium labels on the various fields in the electron tables allows us to search requests, to convert the electron table in the bioresource network centers, international databases, GBIF resources (gbif.org) and conduct statistical data processing. The metadata analysis of herbarium labels was conducted in the following 3 modules: taxonomic analysis (including nomenclatural analysis of the species biodiversity of the families in NSK and NS collections), geography analysis (including location-based analysis of the places of collecting herbarium) and historical analysis (analysis date of collection and collectors). The CSBG Herbarium taxonomic database block is compatible with the international resource "Catalogue of Life" (http://www.catalogueoflife.org), where information is updated every month and an annual list of taxa is published [12].

3.1 Taxonomic Analysis

Total numbers of digitized taxa are 106 families, 443 genera and 1577 species. The largest number of samples scanned are from families Primulaceae (4984), Cystopteridaceae (891), Orchidaceae (750), Boraginaceae (664), Poaceae (632), Asteraceae (550), Athyriaceae (436), Fabaceae (418), Amaryllidaceae (381), Plumbaginaceae (365).

The largest number of samples scanned are from genera Primula (2129), Androsace (1733), Cystopteris (495), Gymnocarpium (396), Allium (376), Stipa (365), Cypripedium (364), Trientalis (342), Lappula (327), Lysimachia (302), Naumburgia (251), Oxytropis (231), Diplazium (224), Athyrium (212), Dryopteris (204), Aconitum (201), Krascheninnikovia (180), Glaux (168), Iris (162), Goniolimon (154), Limonium (150), Neottianthe (146), Cryptogramma (129), Erythronium (121) and Orchis (112).

Special attention was paid to rare and endangered species from Siberia. We digitized 2695 specimens of rare and endangered species from 34 families, 64 genera, 99 species.

3.2 Geography Analysis

Initially, the coordinates were given on 18.9% of herbarium specimens and 69.9% were found online using Google and Yandex search systems. In NSK and NS collections there are samples from 36 countries, including Russia (12013), Kazakhstan (409), Mongolia (158), United States (109), Tadzhikistan (82), Canada (35), Uzbekistan (34), Ukraine (33), Kirgizia (220 and Georgia (22). Russia is represented by 62 subjects of the federation, including completely covered regions of Siberia and the Far East. The following regions are most representative: Irkutsk region (1641), Republic of Buryatia

(1636), Tyva Republic (1621), Krasnoyarskiy kray (1608), Altay Republic (1332), Yakutia (831), Novosibirsk region (592), Khakassia (561), Zabaikalskiy kray (537), Altayskiy kray (265), Kamchatskiy kray (190) and etc.

3.3 Historical Analysis

The earliest herbarium specimens in NSK collection was dated 1829. This sheet is Primula mattiolii subsp. turkestanica (Losinsk.) Kovt., signed by N. Turchaninov "Ad rivulam Bezineme; prope thermes Turkensis", (NSK0009572). Then, the oldest herbarium specimen by K. F. Ledebour, 1833-08-06, Glycyrrhiza uralensis Fisch. and Lappula duplicicarpa Pavlov from Dzhungaria by G. S. Karelin and I. P. Kirillov, 1840–1841.

NS collection stores samples of Lysimachia pentapetala Bunge and Androsace umbellata (Lour.) Merr. since 1851 (Tatarinov, China). The 19th century is represented by 68 sheets in total, including the collections of G. S. Karelin, I. P. Kirillov, A. Regel, P. N. Krylov, R. F. Niman, A. I. Kytmanov and etc.

Total number of herbarium specimens collectors is about 1800. The main collectors in NSK herbarium: M. M. Ivanova, N. S. Vodopyanova, L. I. Malyshev, G. A. Peshkova, I. N. Pospelov, L. V. Bardunov, N. M. Bolshakov, N. V.Vlasova, V. M. Doronkin, A. A. Kiseleva, N. K. Kovonyuk, A. Yu. Korolyuk, N. N. Lashinskiy, S. V. Ovchinnikova, Yu. N. Petrochenko, N. V. Friesen and A. A. Chepurnov.

In NS herbarium the main collectors are: I. M. Krasnoborov, V. M. Hanminchun, D. N. Shaulo, I. A. Artemov, M. P. Danilov, E. I. Korotkova, A. A. Krasnikov, A. V. Kuminiva, M. N. Lomonosova, I. Ya. Neifeld, K. A. Sobolevskaya and S. A. Timokhina.

4 Conclusion

Digital collections became an urgent necessity at the present stage of studying biodiversity and generation of network bioresource centers. The creation of a Virtual herbarium is a modern trend in Botany, the inventory and modernization of herbarium collections of the world's leading botanical institutions.

Specialized programs of Microtek ScanWizard-Botany and MiVapp-Botany greatly facilitate the work on automatic recognition of label text and unique barcode of the herbarium sample, accelerate the introduction of information into xml files, and also create databases with preview images of herbarium sheets.

Acknowledgements. We would like to thank S. N. Lukyanchikov (CSBG SB RAS) and Microtek's engineers for the consultations, L. Z. Lukmanova, S. A. Krasnikova, and I. M. Deyun for the technical support of our research work. The study was made with partial financial support by the Russian Foundation for Basic Research (grant № 15-29-02429) and the base project VI.52.1.5. (state reg. AAAA-A17-117012610055-3).

References

1. Linnaeus, C.: Species Plantarum. 1st edn. Holmiae (1753)
2. Christenhusz, M.J.M., Byng, J.W.: The number of known plant species in the world and its annual increase. Phytotaxa **261**, 201–217 (2016)
3. James, S.A., Soltis, P.S., Belbin, L., Chapman, A.D., Nelson, G., Paul, D.L., Collins, M.: Herbarium data: global biodiversity and societal botanical needs for novel research. Appl. Plant Sci. **6**(2), e1024 (2018). https://doi.org/10.1002/aps3.1024
4. Kovtonyuk, N.K. Virtual collection of type specimens in M.G. Popov Herbarium (NSK). Rast. Mir Asian. Russia **3**(19), 88–93 (2015)
5. Kovtonyuk, N.K.: Virtual herbarium collections as a resource for taxonomy and biodiversity study. Rast. Mir Asian. Russia **1**(25), 98–104 (2017)
6. Kislov, D.E., Bakalin, V.A., Pimenova, E.A., Verkholat, V.P., Krestov, P.V.: An electronic management system for a digital herbarium: development and future prospects. Botanica Pacifica **2**, 59–68 (2017)
7. Seregin, A.: The largest digital herbarium in Russia is now available online. Taxon **67**(2), 463–467 (2018)
8. Doronkin, V.M., Shaulo, D.N., Banaev, E.V., Naumenko, Yu.V: The herbarium collections in the Central Siberian Botanical garden, SB RAS—status and prospects. In: Novikova, L.A. (ed.) Botanical Collections—The National Heritage of Russia 2015, pp. 228–232. PSU, Penza (2015)
9. Kovtonyuk, N., Belyaeva, I. Nomenclatural and taxonomic notes on the names published by M.G. Popov in *Salix* L. and *Populus* L. (Salicaceae). Skvortsovia **2**(2), 126–140 (2015)
10. Han, I.V. Taxa of Asteraceae, Euphorbiaceae and Scrophulariaceae in collections of types of M.G. Popov Herbarium (NSK). Rast. Mir Asian. Russia **3**(23), 41–54 (2016)
11. URL: http://www.microtek.com/products.php?KindID=12&ID=373. Last accessed 07 Aug 2018
12. Roskov, Y., Abucay, L., Orrell, T., Nicolson, D., Bailly, N., Kirk, P.M., Bourgoin, T., DeWalt, R.E., Decock, W., De Wever, A., Nieukerken, E. van, Zarucchi, J., Penev, L. (Eds.) Species 2000 & ITIS Catalogue of Life. 2017, Annual Checklist. Digital resource at www.catalogueoflife.org/annual-checklist/2017. Species 2000: Naturalis, Leiden, the Netherlands. ISSN 2405-884X. http://www.catalogueoflife.org. Last accessed 07 Aug 2018

Taxonomic and Phytogeograpical Databases in Systematics of the Flowering Plant Family Umbelliferae/Apiaceae

Michael G. Pimenov[1](✉), Michael V. Leonov[2], and Tatiana A. Ostroumova[1]

[1] Botanical Garden, Lomonosov Moscow State University, Moscow 119991, Russian Federation
mgpimenov@mail.ru
[2] Faculty of Computational Mathematics and Cybernetics, Lomonosov Moscow State Universty, Moscow 119991, Russian Federation

Abstract. Since the 1980s several databases have been compiled in the Botanical Garden of Moscow State University in collaboration with the Faculty of Computational Mathematics and Cybernetics and the Computer Centre of the same university. These databases deal with various aspects of the systematics and geography of Umbelliferae, one of the largest and most complicated angiosperm plant families. The main goal is to incorporate, store, and retrieve large amounts of information related to genera and species, including scientific names, authors, synonyms, typification, characters, chromosome numbers, distribution, and references to classical and modern publications; and to help prepare monographs, revisions, and articles on various aspects of the family systematics. Placement of most databases on the Internet was beyond the scope of our work, due to limited possibilities. The following databases have been developed–GNOM (Generic NOMenclator: on nomenclature, synonymy, and distribution of world Umbelliferae genera), CARUM (CARyologia UMbelliferarum: on chromosome numbers and karyotypes of world Umbelliferae species), and ASIUM (ASIatic UMbelliferae: on the systematics and geography of the Umbelliferae genera and species of Asia. The monographs *The Genera of the Umbelliferae. A Nomenclator* (based on GNOM; Pimenov & Leonov; Kew, 1993) and "The Karyotaxonomic Analysis on the Umbelliferae" (based on CARUM; Pimenov, Vasil'eva, Leonov & Daushkevich; Enfield, NH, 2002) were published. All three databases are widely used in current work with the prospect of electronic publications. For the monograph *The Umbelliferae of Russia* (Pimenov & Ostroumova; Moscow, 2012) a computer key based on diagnostic character database was compiled.

Keywords: Databases · Taxonomy · Phytogeography · Karyology · Flora · Umbelliferae · Apiaceae · Asia

© Springer Nature Switzerland AG 2019
I. Bychkov and V. Voronin (Eds.): *Information Technologies in the Research of Biodiversity*, SPEES, pp. 28–36, 2019.
https://doi.org/10.1007/978-3-030-11720-7_5

1   Introduction

Documenting plant diversity is now one of the main tasks of any botanical community, being applicable in plant taxonomy, geography, ecology, conservation, and protection, as well as applied botany. The use of databases is one of the most effective ways to register taxa and their distribution. Taxonomic databases are of key significance in the general issue of biodiversity information, in particular in plant systematics, especially in large scale researches of big polymorphic taxa (e.g. families and further), revised on a broad geographical - even global - basis and studied in terms of integrative taxonomy.

The first phytotaxonomic databases appeared at the beginning of the 1970s. They covered grasses (Poaceae) of Australia [1] and Fabaceae, first the tribe Vicieae, then a whole family ([2–4]). Information on these pioneer studies and publications can be found in the monograph "Databases in systematics" [5] and some other reviews [6–8].

The family Umbelliferae (Apiaceae, carrot family, зонтичные) is a taxon, now under critical study in various countries. It has been a traditional object for botanical investigations at Moscow University since the beginning of the 19th century when a German-Moscow botanist Prof. Georg Franz Hoffmann [9, 10], the first University professor of botany and founder of the scientific Botanical Garden, wrote the first monograph on Umbelliferae genera. This has become a tradition. Since the 1970s, the family has been a matter of intensive research in the Botanical Garden at Moscow State University, carried out with the help of the so called multidisciplinary approach, including taxonomy, nomenclature, phylogeny, morphology, micromorphology, anatomy (especially carpoanatomy), phytochemistry, molecular systematics, as well as phytogeography and ecology. Besides numerous particular revisions of genera, mainly Asian, several accompanying summaries [11, 12], some regional treatments were published on the Umbelliferae–of Middle Asia [13], Tajikistan [14], the Russian Far East [15], Siberia [16], Kirghizia [17], and Russia [18]. Participation in such international projects as *Flora Iranica*, *Flora of China* and *Families and Genera of Flowering Plants* should also be mentioned. We have been observing and collecting Umbelliferae in various regions and countries of Asia since 1960. Plants of numerous species have been grown in the Botanical Garden of Moscow State University. Geographically, the research program is focused on Asia (all countries) and the Mediterranean, both being the main diversity centres of the world Umbelliferae.

2   The Botanical Databases Carried Out at Moscow University

The above mentioned early survey works were based on quite large amounts of information, which, however, stored on paper carriers (card files). The necessity of taxonomic and phytogeographical databasing in this work has been evident from the very beginning, even before the era of personal computers and the Internet. Although very useful, those mainly not specialized plant taxonomy databases such as [19–24] did not completely suit our purposes. They did not allow to observe all details of nomenclature and typification, essential for Umbelliferae systematics with its unstable generic concept and resulting extensive synonymy. In the most comprehensive information system "Catalogue of Life", always available online, plant name synonymy borrowed from some authoritative Floras is rather limited and sometimes inconsistent. Using university facilities and close contacts with its scientists from various fields, some databases on different aspects of taxonomy and geography of the Umbelliferae were compiled by experts from the Moscow State University Botanical Garden together with the Faculty of Computational Mathematics and Cybernetics (databases GNOM, CARUM, ASIUM). These databases are like living monographs on various aspects of Umbelliferae systematics.

During last decades more botanical taxonomic databases appeared in Moscow University; some completed (with new data added and retrieved), others in progress, or only at the early stage. The history of database technology used in our project may be of interest, since it mirrors the international IT and programming from the late 1980s until the early 21st century.

3   Databases Published in the Past as Monographs (GNOM and CARUM)

The first one was GNOM (Generis NOmenclator), a database on taxonomy and distribution of the world Umbelliferae. It served as the basis for the monograph "The Genera of the Umbelliferae. A Nomenclator" [25] (Pimenov, Leonov, 1993) published in Kew (the U.K). The main part of the monograph (nomenclator) was generated from the database, first for domestic botany. Strictly speaking, we developed not just a database, but information systems with a developed user interface, the possibility of making of predefined, as well as dynamically generated queries, the most important of which was the output of nomenclature summaries for articles and monographs. The database contains data on the name of the genus, the author(s) of the generic name or a combination, the date and place of protologue publication, typification, references to lectotypification, synonyms (with the protologue and typification data), geographical distribution, the number of species and relevant literature. A menu of queries for the database enables one to get answers to many systematic and floristic questions, for example–to obtain a list of genera described by a particular taxonomist, at a certain time, a list of genera for some geographical area (a continent, a country etc.), a list of nontypified taxa, and a list of genera from each of traditional subfamilies (Hydrocotyloideae, Saniculoideae or Apioideae), tribes or subtribes. The main advantage of the database is that it is an moving monograph, readily reflecting current nomenclature modifications resulting from recent critical revisions, At the same time, however, it presents our personal concept of optimal nomenclature and synonymy of Umbelliferae genera. The first version of GNOM was operated on the EU-1045 under the network DBMS, developed at the MSU Faculty of Computation Mathematics and Cybernetics, and working with data on punch cards. It was also operated on a personal computer Yamaha, manufactured in Japan, using the adapted version of DBase II database (Novosibirsk).

Up to now, the Nomenclator is widely used as a reference system for generic nomenclature and taxonomy of the family. Our later databases are connected to GNOM, thus ensuring the use of only accepted generic names.

The second completed Umbelliferae database, concurrently developed by the same research team, was named CARUM (CARyologia UMbelliferarum). This is a computer database on karyotaxonomy, a special branch in biodiversity studies, that deals with the taxonomic diversity of chromosome number and karyotype morphology. The database contains information about haploid and diploid chromosome numbers, as well as references to karyotype morphology of Umbelliferae species all over the world, with the origin of the studied materials. Karyological data on each studied species are arranged chronologically. The menu of queries of CARUM database enables one to receive analytical answers to various requests concerning the distribution of chromosome numbers within subfamilies, tribes, subtribes, and genera, or chromosome numbers of species from a certain area, or a species, mentioned in this or that publication etc. When published [26] the database CARUM contained data on chromosome numbers and karyotypes of 1,750 species of the world Umbelliferae. The early life of CARUM was the same as that of GNOM. Then, after personal computers became available, both databases were reprogrammed for DBase III. Then a program was developed to convert data on punch cards to an array of batch input, intended already for PC information systems. Thanks to the experience gained with punch cards, we developed batch input for all subsequent systems together with interactive input. Not only did this option significantly speed up the input process, but it also facilitated the recovery of restore the databases in case of failure as well changing the database structure and switch on to new computers. After we had access to personal computers like IBM PC and IBM PC/AT, we moved on to DBase III Plus, KARAT (a russified Foxbase clone), and finally to a very convenient DBMS FoxPro.

During this development period, new botanical tasks were arising, and the challenge was the permanent upgrade of software components to include these tasks in the already prepared databases. Or, in the programming jargon, we repeatedly faced the so-called task of legacy databases. The method of software development itself was applied to prototyping, or rapid prototype.

4   A Database Compiled and Ready for Publication, but Thus Far not Published

Currently under development the ASIUM database is described here in greater detail. It is a system for storage, retrieval and analysis of information on Asian Umbelliferae. Structurally, the ASIUM database consists of related files, which correspond to the principal objects of taxonomic study. These include the files of genera, species and infraspecific taxa, as well as their synonyms, files of type materials, literature references, author's names and their combinations, geographical distribution at three levels of administrative division (countries, regions, and districts) and, alternatively floristic regions and continents. A standard output file contains the accepted species name, authorship, data of the first publication, type locality according to the protologue, type specimen(s) and their localization in herbaria, both already known or newly found, references to essential published Floras, monographs and critical articles; synonyms (with the same set of data as for the accepted name), geographical distribution (countries, as well as parts of countries (regions) and provinces for large countries), geographical distribution in big phytochories in Asia, overall distribution (in other continents). The ASIUM software, the related retrieval system, and the set of subsidiary utilities were initially developed with DBMS FoxPro for DOS, and then transferred to and further developed under Windows (95, 98, 2000, and XP). In order to use modern facilities for information retrieval, a database model was developed in the XML language, together with ASIUM convertors, a set of related XML documents, and programs to deal with them. In particular, an XMLBiblio program was created, allowing for the retrieval and editing of references in UNICODE with the use of diacritic symbols.

The ASIUM information system, in the course of its long life, required some new experimental solutions. Since the output of taxonomic summaries in the system was focused on encoding the MS DOS operating system, and the publications needed to be formatted in the Microsoft Word program, the "legacy database" was used as follows. A utility was written in the FoxPro language that inserted auxiliary symbols, like $ and #, to mark special character design (italics for Latin names and collectors, boldface for accepted names etc.). Additionally, MS Word processor macros were written to convert those words marked with symbols to the necessary font. Thus, we succeeded in minimizing manual labor in the preparation of taxonomic reports.

Apart from standard options, the ASIUM information system has a set of queries most interesting for botanists: the calculation of the number of species and genera for regions of any level, the evaluation of similarity in Umbelliferae sets for any two regions, and the counting of endemics.

The choice of the accepted name for each species reflects our personal opinion, while proceeding from particular investigations. Genera, to which the species are attributed, on the whole correspond to our reference-book *The genera of the Umbelliferae. A Nomenclator* [25], with further additions and corrections made in the working database GNOM. The latter is connected to ASIUM; generic names, absent in GNOM, cannot be used in ASIUM, with rare exceptions of Umbelliferae species erroneously described in other families (*Valerianella, Euphorbia, Eranthis, Chrysosplenium, Limnanthemum, Ranunculus, Ophiorrhiza*, and *Geophila*).

The abbreviations or full transcriptions of the authorities of species and genera comply with the international standard – *The authors of plant names* by BRUMMITT & POWELL [27], except only Chinese botanists' names. Together with [28], we believe that the abbreviations of the names of the Chinese authors of plant names, in the Western style (first initials and surnames) to be of little use; there are too many namesakes, even among botanists, in contrast with the diversity of given names. Therefore, Chinese personal names are used in their full form both in taxa authorities and in the reference list, so as to avoid any misunderstanding.

Abbreviations of periodicals in protologues are made according to the *Botanico-Periodicum-Huntianum* [29], and books after [30], if the bibliographical sources are included in these reference-books. In all other cases we proceeded from the database IPNI [19]. Original Cyrillic titles are transliterated to English using Latin letters; the titles are also translated into English, while some references are accompanied by Latin or German translations of original publications. Acronyms of herbaria correspond to [31].

The search of type materials was one of the most laborious and time-consuming activities when preparing the account. The challenges faced were legion; however, the proposed typification cannot be regarded as complete. There seems to be no end to this. At the first stages of the search for a potential type material, databases of various herbaria, available on the Internet, the so-called *virtual herbaria*, were widely used. It was both the global database of JSTOR (https://plants.jstor.org) and its partner herbaria as FI (Firenze, Museo di Storia Naturale dell'Università, FI, FT, FW; https://plants.jstor.org/partner/FI) and MPU (Montpellier, Herbier de l'Université Montpellier II, MPU; https://plants.jstor.org/partner/MPU), as well as individual databases of large herbaria of the world. Among the last are P (Paris, Muséum Nationale d'Histoire Naturelle; https://www.mnhn.fr/fr/collections/ensembles-collections/botanique/plantes-vasculaires), BM (London, Natural History Museum, (www.nhm.ac.uk), K (London, Royal Botanical Gardens Kew; apps.kew.org/herbcat/navigator.do), G (Genève, Conservaroire et Jardin Botaniques Genève; http://www.ville-ge.ch/musinfo/bd/cjb/chg/), W & WU (Wien, Naturhistorisches Museum & Universität Wien; http://herbarium.univie.ac.at), E (Royal Botanic Garden Edinburgh; http://data.rbge.org.uk/search/herbarium/), Moscow, (Moscow State University herbarium, MW; https://plant.depo.msu.ru/), B (Berlin, Botanischer Garten und Botanisches Museum, Berlin-Dahlem; https://www.bgbm.org/en/herbarium), S (Stockholm, Naturhisoriska riksmuseet; http://herbarium.nrm.se/search/species/), L & WAG (Leiden, Naturalis Biodiversity Center; https://science.naturalis.nl/en/collection/naturalis-collections/botany), GOET (Göttingen, Georg-August Universität; https://www.uni-goettingen.de/de/185941.html), M (München, Botanische Staatsammlung München; www.botanischestattsammlung.de/general/herbarium.html); MSB (München, Ludwig-Maximilians-Universität; http:botanik.biologie.uni-muenchen.de/botsyst/), O (Oslo, Natural History Museum, Botanical Museum, University of Oslo; https://www.nhm.uio.no/fagene/botanikk), UC (Berkeley, California University Herbarium; https://webapps.cspace.berkeley.edu/ucjeps/publicsearch/publicsearch/), GB (Göteborg, University of Gothenburg, Herbarium, Dept. of Biological and Environmental Sciences; http://bioenv.gu.se/forskning/forskningsresurser/herbarium)

Then Umbelliferae collections have been studied in Russia and abroad, including famous centers of taxonomic botany as Kew, Geneve, St.Petersburg, Edinburgh, Vienna, Berlin-Dahlem, British Museum, Moscow, Paris, in numerous local herbaria in China, Turkey, Iran, India, and Nepal; some essential materials were obtained on loan by exchange. The full list of herbaria, whose materials were studied, is quite impressive, to include: A, AA, ANK, ASH, B, BAK, BEY, BLAT, BM, BP, C, CAL, CALI, CDBI, DD, E, ERE, FI, FRU, G, GB, HUB, HUJ, ISTE, IZM, JE, K, KATH, KW, KUN, KYO, L, LE, LINN, LIV, MANCH, M, MAK, MH, MHA, MPU, MW, NAS, NS, OXF, P, PE, TAD, TAK, TARI, TBI, TI, TK, UPS, VAN, VLA, W, WU, XJBI abbreviated according to [31]. We are grateful to the curators and staff of listed herbaria for the opportunity to examine the materials. The latest contribution is the addition to ASIUM of barcodes of type specimens kept in various herbaria (if existing).

A special option allows the arrangement of synonyms, which appear to be abundant in many Umbelliferae species, due to constant changes in generic nomenclature. A new term – *homotypic group* of synonyms – has been introduced. A homotypic group is formed by names, based on the same type material. These names are also called nomenclatural synonyms. Names, based on different types (taxonomical or heterotypic synonyms) form different homotypic groups. Misapplied names are out of either group. Synonyms have been sorted manually, with the help of a special option, since there are no formal characters (homotypic synonyms might not have identical epithets).

References to botanical literature for each species' name have been made in the chronological order and in the abbreviated form – author(s), year of publication, and page number. Full bibliographical data are added to the reference list, which seems to be particularly useful with the Asian Umbelliferae bibliography. The database contains over 2,300 references, mainly on local Floras, monographs, and critical articles.

ASIUM is a living nomenclatural and phytogeographical monograph, open for additions and corrections. One can use its information to make various comparisons, including species lists of various territories, to obtain floristic similarity measures, to compile lists of endemic species, to produce lists of type specimens, kept in particular herbarium, etc. Intermediate results of registration of Asian Umbelliferae diversity were published [32–35].

ASIUM is a database, which accumulates massive data of various kinds, including 1. results of our own taxonomic and floristic investigations, made in different parts of Asia; 2. data from various herbaria, with special emphasis on type collections; and 3. data from published Floras and critical articles on taxonomy, phylogeny and phyto-geography of Asian species.

5   Identification Keys

Umbellliferae belongs to the most complicated families of flowering plants, also in terms of species identification. Traditional printed dichotomous keys in large "Floras" are far from satisfying. They often demand leaf, flower and fruit characters, and the absence of petals or mature fruits makes identification impossible. Some important characters, e.g. leaf dissection, can hardly be described, and the keys often contain characters which are difficult to observe, e.g. small secretory ducts visible only on fruit transections at magnification x400. The way out this problem could be computer-assisted identification, based on descriptive data banks [36]. The first keys of this type appeared together with the first botanical databases; the early steps were described in the book *Biological Identification with Computers*, the proceedings of the 1973 Cambridge conference, edited by Pankhurst [37]. Interactive multi-entry polytomous keys with illustrations for personal computers are the most convenient instruments [36]. We developed software and compiled the key for Russian Umbelliferae (288 species) based on 55 structural characters and geographical distribution [38]. Character states are illustrated and thus can be used by the inexperienced people as well. The list of character states can easily be translated into any language; we have created Russian, English, German, and French versions.

Users begin identification with the most convenient characters of the specimen. After a few steps, they will have a short list of species with a chosen combination of character states, growing in a selected region. Our key contains photos of herbarium specimens for all the Russian Umbelliferae, so identification could easily be verified. An advantage of our key is a table of character states that can be generated for any 1–5 species, with diagnostic characters being highlighted. Users can choose characters for further identification or check their results based on the overall list of characters.

The most labour-intensive part of creating keys is compiling the table of characters (descriptive data bank), in our case, 55 characters for 288 species. Gaps in the table make "noise" during testing and identification, thus decreasing the accuracy, so we had to put a lot of effort into minimizing the number of empty cells. The published descriptions differ in details and inconsistent terminology, therefore most of the data were checked on herbarium specimens, living plants and microscopic slides. The necessity of collecting a large body of data seems to be one reason why computer keys have not thus far been compiled for most groups of organisms.

References

1. Watson, L., Dallwitz, M.J.: Australian Grass Genera: Anatomy, Morphology and Keys. Australian National University Canberra (1980)
2. ILDIS (International Legume Database and Information Service). https://www.ildis.org
3. Bisby, F.A., White, R.J., Macfarlane T.D., Babac V.T.: Vicieae database project: experimental uses of the monographic taxonomic database for species of vetch and pea. In: Felsenstein, J. (ed.) Numerical Taxonomy. Proceedings of the NATO Advances Study Institute Bad Windsheim, 4–6 July 1982, pp. 625–629. Springer, Berlin (1983).
4. Bisby, F.A.: Automated taxonomic information systems. In: Heywood, V.H., Moore, D.M. (eds.) Current Concepts in Plant Taxonomy. The Systematics Association, special vol. 25. Academic Press, London (1984)
5. Allkin, R., Bisby, F.A. (eds.): Databases in Systematics. The Systematics Association, special vol. 26. Academic Press, London (1984).
6. Abbott, L.A., Bisby, F.A., Rogers, D.J. (eds.): Taxonomic Analysis in Biology. Computers, Models, and Databases. Columbia University Press, New York (1985)
7. Pimenov M.G.: Mathematical methods and computing technique in higher plant systematics. VINITI, Itogi Nauki i Tehniki (Outcomes in Science and Technique). Botany, vol. 8, no. 2. VINITI Press, Moscow (1988) (in Russian).
8. Pankhurst, R.J.: Practicing Taxonomic Computing. Cambridge University Press, Cambridge (1991)
9. Hoffmann, G.F.: Genera plantarum Umbelliferarum eorumque characters naturales secundum numerum, figuram, situm et proportionem fructificationis et fructus partium. Typis N.S. Vsevolozskianus, Mosquae (1814)
10. Hoffmann, G.F.: *Plantarum Umbelliferarum genera. Ed. 2* Typis N.S.Vsevolozskianus, Mosquae (1816)
11. Pimenov, M.G., Tikhomirov, V.N.: Typificatio generum Umbelliferarum florae URSS. Novosti Sistematiki Vysshykh Rastenij **16**, 154–166 (1979) (in Russian, Latin)
12. Pimenov, M.G., Constance, L.: Nomenclature of supraganeric taxa in Umbelliferae/Apiaceae. Taxon **34**(3), 493–501 (1985)
13. Pimenov, M.G.: Umbelliferae. In: Vvedensky, A.I. (ed.) Conspectus florae Asiae Mediae. vol. 7, pp. 167–322. FAN, Tashkent (1983) (in Russian)
14. Korovin, E.P., Pimenov, M.G., Kinzikaeva, G.K.: Umbelliferae. In: Ovczinnikov, P.N. (ed.) Flora Tajikskoi SSR, vol. 7, pp. 10–214. Akad. Nauk SSSR, Leningrad (1984) (in Russian)

15. Pimenov, M.G.: Umbelliferae. In: Kharkevish, S.S. (ed.), Sosudistye rastenija Sovetskogo [Rossiyskogo] Dal'nego Vostoka [Vascular plants of the Soviet Far East], vol. 2, pp. 203–277. Leningrad, Nauka (1987) (in Russian).
16. Pimenov, M.G.: Apiaceae (Umbelliferae) In: Peschkova, G.A. (ed.) Flora Sibiri, vol. 10, pp. 123–194. Nauka, Novosibirsk (1996) (in Russian)
17. Pimenov, M.G., Kljuykov, E.V.: Zontichnye (Umbelliferae) Kirgizii [The Umbelliferae of Kirghyzia]. KMK Scientific Press, Moscow (2002) (in Russian)
18. Pimenov, M.G., Ostroumova, T.A: Zontichnye (Umbelliferae) Rossii (Umbelliferae of Russia). KMK Scientific Press, Moscow (2012) (in Russian)
19. The International Plant Names Index (IPNI). http://www.ipni.org/
20. Tropicos. www.tropicos.org.
21. ITIS.: The Integrated Taxonomic Information System. https://www.itis.gov/
22. Catalogue of Life. http://www.catalogueoflife.org/
23. The Plant List. A Working List of all Known Plant Species. http://www.theplantlist.org/
24. JSTOR Global plants. https://plants.jstor.org/
25. Pimenov, M.G., Leonov, M.V.: The genera of the Umbelliferae. A nomenclator. Kew: Royal Botanic Gardens (1993)
26. Pimenov, M.G., Vassiljeva, M.G., Leonov, M.V., Daushkevich, Ju, V.: Karyotaxonomical analysis of the Umbelliferae. Science Press, Enfield (2002)
27. Brummitt, R.K., Powell, C.E.: The authors of plant names; a list of authors of scientific names of plants, with recommended forms of their names, including abbreviations. Kew: Royal Botanic gardens (1993)
28. Xu Z., Nicolson, D.H.: Don't abbreviate Chinese names. Taxon **43**(3), 499–504 (1992)
29. Hunt Botanical Library: BPH-2: periodicals with botanical content. (Bridson, G.D.R., comp.) 2nd ed. 2 vols. Hunt Institute for Botanical Documentation, Pittsburg (2004)
30. Stafleu, F.A., Cowan, R.A.: Taxonomic literature: a selective guide to botanical publications and collections with dates, commentaries and types. 7 vols, supplements 1-VI by Stafleu F. A. & E.A. Mennega Scheltema & Holkema, Berlin (1976–2000)
31. Thiers, B.: Index Herbariorum. A global directory of public herbaria and accociated staff. New York Botanical Garden's Virtual Herbarium (continuously updated). http://sweetgum. nybg.org/science/ih/
32. Pimenov, M.G., Leonov, M.V.: The taxonomic databases on the Umbelliferae: the current state. Trudy Zoologičeskogo Instituta Rossiiskoi Akademii Nauk, vol. 278, p. 69. St-Petersburg (1999) (in Russian)
33. Pimenov, M.G., Leonov, M.V.: Asia, the continent with highest Umbelliferae biodiversity. S. Afr. J. Bot **70**(3), 417–419 (2004a)
34. Pimenov, M.G., Leonov, M.V.: The Asian Umbelliferae biodiversity database (ASIUM) with particular reference to South-West Asian taxa. Turk. J. Bot. **28**, 139–145 (2004b)
35. Pimenov, M.G., Leonov, M.V., Ostroumova, T.A.: Computer in service of studies in systematics and phylogeny of the Umbelliferae in Moscow University. In: Botaničeskie Issledovanija v Aziatskoi Rossii, vol. 1, pp. 276–277. Barnaul (2003) (in Russian)
36. Lobanov, A.L., Ryss, A.Y.: Computerized identification systems in zoology and botany: present state and prospects. In: Information retrieval systems in biodiversity research. Proceedings of the Zoology Institute RAS, vol. 278, pp. 20–29. St-Petersburg (1999)
37. Pankhurst, R.J. (ed.): Biological identification with computers. The Systematics Association, special vol. no. 7. Academic Press, London (1975)
38. Ostroumova, T.A., Ostroumov, O.S., Pimenov, M.G.: Illustrations. Identification key. Supplement to the book Umbelliferae of Russia. Moscow, KMK Press. CD ROM (2012)

The Fractal Model of the Microorganism's Frequencies Spectrum for Determining the Diversity of the Biochemical Processes in Soil

N. I. Vorobyov[1]([✉]), V. N. Pishchik[2], Y. V. Pukhalsky[1],
O. V. Sviridova[1], S. V. Zhemyakin[3], and A. M. Semenov[4]

[1] ARRIAM, Podbelsky sh., 3, Pushkin, Saint-Petersburg, Russia
Nik.IvanVorobyov@yandex.ru
[2] AFI, Grazhdansky pr., 14, Saint-Petersburg, Russia
[3] Company "Petersburg's Biotechnologies", Tinkoff lane 7D, Pushkin, Saint-Petersburg, Russia
[4] MSU named Lomonosov MV, The Lenin Hills, 1, Moscow, Russia

Abstract. Modern molecular genetic methods have provided a extended opportunities for studying the frequency taxonomic composition of soil microbial communities, including uncultivated strains of microorganisms. Modern spectrum analysis of operational taxonomic units (OTU) identified by 16Sr RNA gene is limited to the comparison of the nucleotide sequences. We propose to use a fractal model to describe the frequency spectrum of microorganisms. This model necessary to study the network self-organization of microbial communities and the functional diversity of the processes of biological systems. The power series from three frequencies of OTU are the fractal model of frequencies spectrum of OTU (for example: 0.1, 0.01, 0.001). Because there are many OTU's peaks in the original spectrum, peaks grouped with the same or similar amplitudes to form OTU-groups. Fractal topological analysis of the location of OTU-groups on a fractal portrait of the microbial community was been used to search for degenerate triangles with OTU-groups at the vertices, since these OTU-groups correspond to a microbial biosystem and to fractal model for the grouped OTU's frequencies. The functional diversity index (I_{FD}) of biochemical processes is been determined by number of OTU-groups in microbial community and by number of OTU-groups in biosystem. The I_{FD}-dynamics was been calculated with using the daily data of molecular-genetic analysis of soil samples. As a result, we revealed the periodic fluctuations in I_{FD}-values and the periodic replacing of OTU-groups in a soil microbial biosystems that provide a wide diversity of a biochemical processes in soil microbial communities.

Keywords: Fractal model of microorganism's frequency spectrum · Functional diversity of biochemical processes in soil microbial communities · Fractal portrait of microbial community

© Springer Nature Switzerland AG 2019
I. Bychkov and V. Voronin (Eds.): *Information Technologies in the Research of Biodiversity*, SPEES, pp. 37–41, 2019.
https://doi.org/10.1007/978-3-030-11720-7_6

1   The Fractal Model of Frequencies Spectrum of OTU-Groups

As components of biological systems, soil microorganisms with maximum efficiency convert organic substrates into nutrients for plants. To analyze the diversity of bio-chemical processes in microbial biosystems, it is been proposed to use molecular-genetic data of microbial communities. Since there are many peaks in the original frequency spectrum of the operational taxonomic units (OTU) of microbial communities, the peaks were been first grouped with the same or similar amplitudes to form OTU-groups. To group OTU, the entire range of values of $ln(p_i/p_{max})$ is divided into intervals in steps of 0.1, where pi, p_{max} are the OTU-frequencies with the serial number (i) and the highest OTU-frequency. The OTU's frequencies which falling into one interval are been summed to form the common frequency of the OTU-group. As a result, the initial spectrum of the OTU-frequencies can been reduced to 20–40 frequencies of OTU-groups.

In the frequencies OTU-group spectrum, it is necessary to select three OTU-groups, which frequencies corresponding to a power series as the fractal model [1]. We assume that these three OTUs form a biosystem and perform one biochemical function. The fractal model is a power series of three frequencies of OTU-groups (for example, 0.1, 0.01, 0.001). The logarithms of these frequencies vary linearly (for example, $-\ln(10)$, $-2\ln(10)$, $-3\ln(10)$,...). This property of the fractal model was been used to search for the OTU-groups belonging to biosystems.

2   The Topological Analysis of OTU-Groups Location on the Fractal Portrait of the Microbial Community

The fractal portrait of a microbial community is the two-dimensional field with points representing the OTU-groups (See Fig. 1) [2]. The Y-coordinate of the OTU-group determines by $ln(q_i/q_{max})$. The X-coordinate of the OTU-group determines by the fractional part of $ln(q_i/q_{max})$, where q_i, q_{max} are the OTU-group's frequency with the serial number (i) and the highest OTU-group's frequency.

The fractal measure of topological analysis is a triangle with three OTU-groups at the vertices. The smallest height (h_i, see Fig. 1) of the triangle is a scale of the fractal measure. At the vertices of the triangles, there must be OTU-groups for which the integral parts of $ln(q_i/q_{max})$ are not equal. For example, the integral parts of $ln(q_1/q_{max})$, $ln(q_{10}/q_{max})$, $ln(q_{17}/q_{max})$ are equal respectively 0, -1, -3 (see the points A, B, C in Fig. 1). All triangles that constructed on the fractal portrait can been arranged in ascending order of the smallest height of the triangles (h_N, see Fig. 2), where N is the serial number of the triangle in this series.

The logarithmic dependence (see Fig. 2) is a quasilinear dependence and it is equivalent to the classical fractal logarithmic dependence [3]. An unusual way of covering by triangles the points on the fractal portrait has led to the fact that the coefficient of fractal dimension has acquired a negative value (-0.86, see Fig. 2).

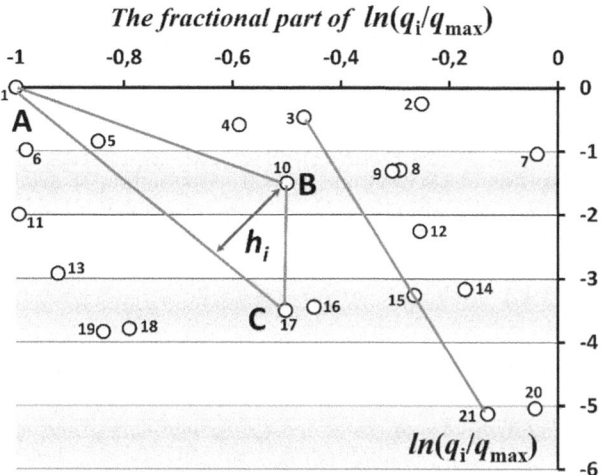

Fig. 1. The topological analysis of OTU-groups location on fractal portrait of the microbial community. Notes: The OTU-group's frequency determines Y and X coordinates corresponding point on the portrait. The ABC-triangle with OTU-groups No. 1, 10, 17 at the vertices is a fractal measure of topological analysis. The smallest height of the triangle (h_i) is a scale of the fractal measure. Since the OTU-groups No. 3, 15, 21 are located on the same straight line and corresponding to a degenerate triangle, we assume that these OTU-groups form a biosystem and perform one biochemical function.

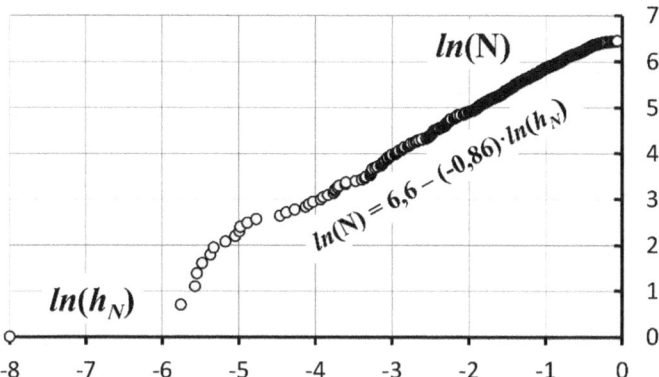

Fig. 2. The logarithmic dependence of the serial number (N) of triangles constructed on a fractal portrait and arranged in ascending order of the smallest height of the triangles (h_N).

Among all triangles that were constructed on the fractal portrait we choose degenerate triangles for which $ln(h_N) < -5$. We believe that the OTU-groups of these triangles form a biosystem and perform one biochemical transformative function. For example, the OTU-groups No. 3, 15, 21 (see Fig. 1) correspond to a degenerate triangle and to a biosystem performing one biochemical function.

3   The Functional Diversity Index of Biochemical Processes in Soil Microbial Biosystem

The number of degenerate triangles found on a fractal portrait determines the number of different processes of biochemical transformation performed by the soil microbial community. The number of OTU-groups (M_B) corresponding to degenerate triangles divided by the total number of OTU-groups in the microbial community (M_C) is the functional diversity index of a biochemical processes in microbial communities (I_{FD}, see Formula (1)).

$$I_{FD} = M_B/M_C. \tag{1}$$

In accordance with the proposed computational method, the daily dynamics of I_{FD} (1) was been calculated with using the molecular-genetic data of soil samples obtained in September 2006 in Moscow Region [4] (see Fig. 3).

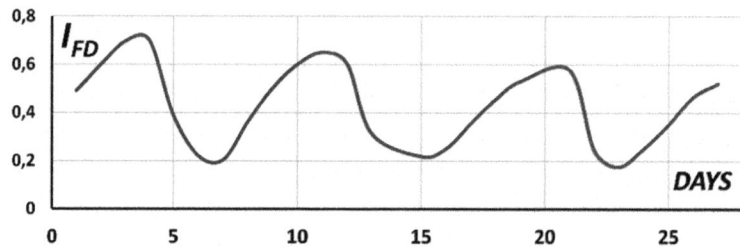

Fig. 3.   The computational daily dynamics of the functional diversity index (I_{FD}) of biochemical processes in microbial communities from the Moscow Region soils [4].

The observed I_{FD} fluctuations (see Fig. 3) mean that in the soil there are constant changes in the structure of microbial biosystems and changes in the direction of biochemical processes. The periodic rearrangement of microbial biosystems with the replacement of OTU-groups and the change in the number of different biochemical transformations points to a wide range of biochemical processes which can supported by the soil microbial community.

Acknowledgements. This work is been partially supported by the Russian National Fund grant No. 14-26-00094.

References

1. Schroeder, M.: Fractals, Chaos, Power Laws: Minutes from an Infinite Paradise. W.H. Freeman, NY (1991)
2. Vorobyov, N.I., Sviridova, O.V., Patyka, N.V., Dumova, V.A., Mazirov, M.A., Kruglov, Y. V.: Fractal-taxonomic portrait of microbial community as bioindicator of a type of soil

destructive processes. In: Book of Abstracts of the International Conference "Bioindication in the Ecological Assessment of Soil and Related Habitats", p. 38. Moscow, BINOM (2013)
3. Mandelbrot, Benoit B.: The Fractal Geometry of Nature. W.H. Freeman and Company, NY (1982)
4. Semenov, A.M., Bubnov, I.A., Semenov, V.M., Semenova, E.V., Zelenev, V.V., Semenova, N.A.: Daily dynamics of bacterial numbers, CO_2 emissions from soil and relationships between their wavelike fluctuations and succession of the microbial community. Eurasian Soil Science, vol. 46(8), pp. 869–884. Pleiades Publishing, Ltd. (2013)

Actualization of Herbarium Labels Information

Olga Pisarenko[1]([✉]), Igor Artemov[1], Sergey Kazanovsky[2],
and Ekaterina Prelovskaya[2]

[1] Central Siberian Botanical Garden SB RAS, Novosibirsk, Russia
{o_pisarenko, artemov_l}@mail.ru
[2] Siberian Institute of Plant Physiology and Biochemistry SB RAS, Irkutsk,
Russia
skazanovsky@mail.ru

Abstract. Date on Russia's bryoflora have a centralized accumulation platform –
the online database (DB) of the "Moss Flora of Russia" (http://arctoa.ru/Flora/
basa.php). The DB intensively develops due to joint effort of currently working
Russian bryologists. But data of vast old herbarium collections remains inaccessible, digitizing of the information requires huge labor costs. To facilitate and
speed up the process desktop database of moss specimen labels is created in the MS
Access. It consists of three tables, queries and a form for data entry; the structure
and ability of the DB are described. The interface of the DB uses fields with drop-
down lists to enter the main part of text data. An important and convenient feature
of the software is implementation of procedure for updating lists of locations,
habitats and collectors. They are replenished in parallel with filling of the DB and
updated every time you enter a new record. The resulting table is convenient for
converting data into DB "Moss Flora of Russia". Now the software is used for
inventory of the old non-arranged moss collections of NSK and IRK.

Keywords: Moss virtual collection database siberia

1 Introduction

Electronic databases keeping information of herbarium samples or records on species
finds, are valuable resources on biodiversity and biogeography. They are both relatively
small table databases of individual researchers and internet-available large virtual
collections (Tropicos, GBIF, Sweden's Virtual Herbarium and others). In the first case
databases facilitate the work of researchers with their own material on distribution,
ecology or polymorphism of species. In the second case databases are the basis for
revealing and modeling the spatial distribution of species in the world scale. Inter alia,
this information is needed to manage resource species, to protect endangered species or
to prevent the spread of invasive ones.

However, the solution of these tasks directly depends on the completeness of a
database. If designing a database and creating a demo version as a rule easily fits two
years of a standard project, the filling of a database of a large collection in the absence

I. Bychkov and V. Voronin (Eds.): *Information Technologies
in the Research of Biodiversity*, SPEES, pp. 42–47, 2019.
https://doi.org/10.1007/978-3-030-11720-7_7

of sufficient funding and staff may be delayed indefinitely. The most time consuming process when creating a database is typing label texts. Thereby, it is important to accelerate the input of sample metadata. Several solutions can be proposed for the purpose.

One of options is scanning of samples and creation of virtual collections. In this case, the information from labels is available as a graphic image. This allows to reduce the typed metadata of samples to a minimum. Thus, when forming the virtual herbarium of Moscow State University, at the initial stage of data processing, the set of sample metadata is reduced to the name of the taxon, under which the sample is stored in the herbarium, the barcode number and the index of one of 60 herbarium regions [11]. However, such reduction of metadata limits the database capabilities for subject search of samples to solve problems in botanical geography and ecology. It is worth noting that in the virtual herbarium of Moscow State University further input of label data is enabled. This work involves filling a MS Excel table from a PC keyboard, that is carried out by operators and volunteers among the interested experts [3]. Information of almost 110000 labels was entered in the virtual herbarium by the middle of 2018, that amounts one-eighth of the total number of scanned samples [8], thus demonstrating efficiency of extensive methods. To speed up the input of information from herbarium labels into a database, it is also possible to use computer technologies for voice and text recognition [9]. But the text obtained as a result of recognized speech always requires editing (arrangement of punctuation, capital letters in toponyms, hyphens in compound words and so on). The recognized text of scanned labels is also very often needs editing. It should be noted that the labels of old samples are most often written by hand and they have to be entered into database manually. In addition, recognized and corrected label text needs to be filled into appropriate fields of the database table. Therefore, the gain in time when using voice or text recognition systems is unlikely to be significant and justified.

As a reasonable alternative to typing the text of herbarium labels on the keyboard or its recognition, it is possible to use desktop databases with a specialized user interface for entering the metadata of herbarium samples. To speed up data entry and eliminate errors, there are some options which are successfully applied in various database management systems: automatic and sequential moving cursor by fields, selecting appropriate records from a drop down lists, automatically updating of lists in the replenishing of a database, export of tables.

Such options can be useful for both creating small desktop databases and preparing data for export to multi-user online databases, for example, such as the database "Moss Flora of Russia". It is a centralized platform for collecting information about moss finds [4, 5]. The database is being intensively developed and presently includes more than 130000 records. It is created by combining the data provided by currently working Russian bryologists. But in the database there are almost no materials from considerable collections of already deceased bryologists: L.V. Bardunov (IRK), V. Ya. Cherdantseva (VLA), A.N. Vasilyev (KRAS), P.N. Krylov and E.Ya. Muldiyarov (TC). The lack of this information biases the general picture of species distribution.

Thus, on maps generated by the "Moss Flora of Russia" DB some regions (the Irkutsk and Tomsk regions, Tuva, Khakassia, the south of the Krasnoyarsk region) appear to be "white spots". For example, for the coordinate intervals 50–55°N and 100–105°E only 95 species are present in the database [5].

However, in reality this area is bryologically relatively well investigated. L.V. Bardunov, S.G. Kazanovsky, N. Dudareva, E.S. Prelovskaya collected mosses in many localities of this territory, which are pointed on the scheme (Fig. 1). The main problem of elimination bryological white spots is not to investigate the territory but to enter the already existing data with minimum labor costs.

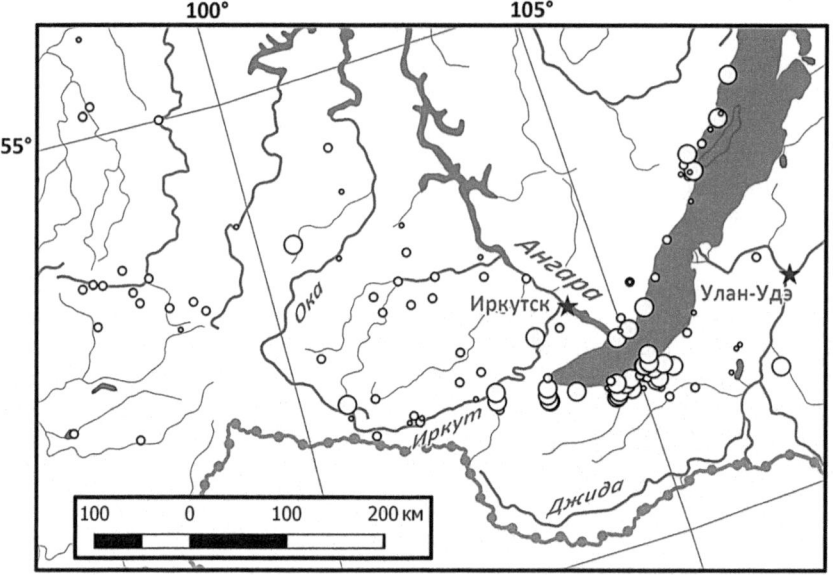

Fig. 1. Areas of bryophyte investigations in the southern Baikal region. Circle size reflects the state of knowledge: the smallest circles – occasional collections; the biggest ones – thoroughly examined localities on Spring 2018 [1, 2, 6, 7, 10].

2   Materials and Methods

A database of moss specimens labels is created in the MS Access. It consists of three tables, queries and a form for data entry.

Fields of the DB metadata table correspond to the main fields of the database "Moss Flora of Russia" [5], that is a prerequisite for successful data export. There are fields for the following information: Record number – Generic name and species epithet– Intraspecific taxon – Taxonomical status – Region of Russia according to the scheme adopted for "Moss Flora of Russia" – General location – Exact location – Habitat – Elevation – Latitude in decimal degree – Longitude in decimal degree – Collection date – Specimen number set by its collector – Collector's name – The name of an expert who identified a species – The name of confirming expert – Sporophyte presence – Admixture – Herbarium code – Herbarium number – Identification history – Geoposition accuracy – Source of coordinates – Comment.

Reference tables are used in the database for the formation of fields with lists: the table of the moss names and the table of region division.

To quick entering the specimens label metadata in the DB, a form with implemented options is used (Fig. 2).

After typing information in the appropriate field, the cursor automatically moves to the next specified field, with corresponding switch of a language.

Some fields are filled in automatically. For example, after choosing the value of "Genius" and "Species" fields, the resulting field with binominal name is completed automatically.

The fields with latitude and longitude of locality in decimal degree filled up as result of automatic recalculation after entering of degree and minutes values in the corresponding fields.

The generic and species name and region code are entered in the fields with drop-down lists "Genius", "Species" and "MFR_region". As a source of data for the lists, there are reference tables of Specific Names and Regions, as well as queries created on their basis. When entering a generic name, one can select it with the cursor in the drop-down list or enter the initial letters of the genus in the field and the desirable element will be selected automatically. After entering the generic name, the value is used as the selection condition for updating the query to the Species names table. The column of the species names of the query is used as a data source for the field with the "Species" list: the last is a list of specific epithets of the entered genus from the reference table. Entering a specific epithet is similar to entering a generic name–by selecting from the list or by typing its initial letters. The region code is entered similarly.

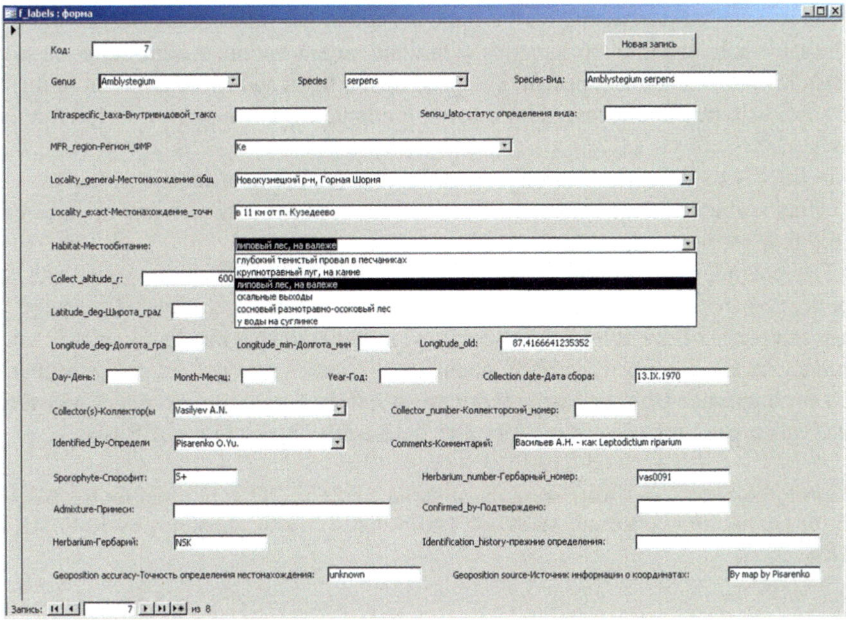

Fig. 2. Form for entering metadata of a herbarium specimen.

Information about the location, habitat, collector name and name of determining expert are entered in the fields with the updated drop-down lists "Locality_general", "Locality_exact", "Habitat", "Collector (s)", "Identified_by". The lists for these fields are not prepared in advance in the form of directories, but are formed on the basis of queries to the Metadata table. They propose the nonrepeating values that were previously entered in the table. A query is updated each time when a new record is entered. If under entering a label it turns out that the metadata table already has records from the same locality or habitat, it needs only to select the required value from the drop-down list (Fig. 2). In the absence of suitable values, it could be entered from the keyboard. Since queries and drop-down lists of fields are updated each time a new value is entered, this value will appear in the list when the next record entering.

The field "Collection date" is organized as a text field formed on the base of the integer data fields "Collection day", "Collection month" and "Year of collection". It is done to speed up the entry of a date. According to the accepted standard, the month is written in Roman numerals. In the proposed solution the "Collection date" field is filled in automatically according to the standard after entering the day, month and year numbers in the corresponding fields.

3   Results and Discussion

The primary information received by researchers about the distribution and ecology of living organisms is a poorly structured array of descriptive data. To involve this information in the analysis, it is necessary to arrange appropriate databases.

The work of biologists and, in particular, botanists in the field has a certain specificity. So, in floristic studies, the same collector, as a rule, collects several samples in the same habitat and many samples within the same location. Accordingly, on many labels, this information is repeated. To speed up the input and avoid errors in filling the database, it is possible to use pre-prepared dictionaries, but that is a fairly large and routine task.

In the discussed moss label database, the reference lists of locations, habitats and collectors are replenished in parallel with the filling of the database and updated each time a new record is entered.

The operation of the DB has shown that these interface options facilitate and significantly speed up digitization of information of herbarium samples. Currently, the moss samples collected by L.V. Bardunov in Siberia and the Far East are under digitising by the described interface; about 1000 samples are now entered. Information from the metadata table is easily exported to external files, for example, MS Excel table, which used for converting data into online-DB "Moss Flora of Russia".

Acknowledgements. This work was carried out in the frame of the program for the development of bio-resource collections (UNU № USU 440537), partly supported by RFFI (18-04-00822).

References

1. Bardunov, L.V.: Mosses of East Sayan. Nauka, Moscow-Leningrad (1965). (In Russ.)
2. Dudareva, N.V.: Bryoflora of East Sayan vicinity. Ph.D. diss. SIPPB, Irkutsk (2006). (In Russ.)
3. Forum "Plantarium". Resources on botany. Digital Herbarium of the Moscow State University. Homepage, http://forum.plantarium.ru/viewtopic.php?id=53604. Last accessed 19 June 2018
4. Herbarium Specimens of Russian Mosses. Homepage, http://arctoa.ru/Flora/basa.php. Last accessed 23 July 2018
5. Ivanov, O.V., Kolesnikova, M.A., Afonina, O.M., Akatova, T.V., Baisheva, E.Z., Belkina, O.A., Bezgodov, A.G., Czernyadjeva, I.V., Dudov, S.V., Fedosov, V.E., Ignatova, E.A., Ivanova, E.I., Kozhin, M.N., Lapshina, E.D., Notov, A.A., Pisarenko, OYu., Popova, N.N., Savchenko, A.N., Teleganova, V.V., Ukrainskaya, GYu., Ignatov, M.S.: The database of the moss flora of Russia. Arctoa **26**, 1–10 (2017)
6. Kazanovsky, S.G.: Bryoflora of Khamar-Daban ridge (South Baikal region). Ph.D. diss. CSBG, Novosibirsk (1993). (In Russ.)
7. Makry, T.V., Kazanovsky, S.G., Bardunov, L.V., Prelovskaya, E.S.: Spore plants of the Baikal national Park. Geo, Novosibirsk (2008). (In Russ.)
8. National Bank-depository of living systems. Digital Herbarium of Moscow State University. Homepage, https://plant.depo.msu.ru/. Last accessed 19 June 2018
9. Nelson, G., Paul, D., Riccardi, G., Mast, A.R.: Five task clusters that enable efficient and effective digitization of biological collections. Zookeys **209**, 19–45 (2012)
10. Prelovskaya, E.S.: Bryoflora of south–west coast of Baikal Lake. Ph.D. diss. BSI, Vladivostok (2010). (In Russ.)
11. Seregin, A.P.: The Moscow University Digital Herbarium—the largest russian biodiversity database. Izvestiya RAN. Ser. Biol. **6**, 30–36 (2017). (In Russ.)

Ground Surveys Versus UAV Photography: The Comparison of Two Tree Crown Mapping Techniques

Maxim Shashkov[1,2(✉)] , Natalya Ivanova[1,2] ,
Vladimir Shanin[1,2,3] , and Pavel Grabarnik[1]

[1] Institute of Physicochemical and Biological Problems in Soil Sciences of the Russian Academy of Science, Institutskaya str., 2, 142290 Pushchino, Russia
Max.carabus@gmail.com
[2] Institute of Mathematical Problems of Biology RAS – The Branch of the Keldysh Institute of Applied Mathematics of the Russian Academy of Sciences, Professor Vitkevich str. 1, 142290 Pushchino, Russia
[3] Center for Forest Ecology and Productivity of the Russian Academy of Sciences, Profsoyuznaya str., 84/32, bld. 14, 117997 Moscow, Russia

Abstract. The aim of this investigation was to compare tree crown area obtained by the classical ground survey methods (GS) and by the unmanned aerial vehicle (UAV) technology. The study was carried out in Prioksko-Terrasny Biosphere Natural Reserve (Moscow region, Russia) on permanent sampling plot of 1 ha. All trees with DBH at least of 5 cm (779 ind.) were mapped and measured during ground survey in 2016. For each alive tree the radii of crown horizontal projection in four cardinal directions were measured, and the total area of all projections was calculated. The aerial photography by a quadcopter DJI Phantom 4 was conducted in 2017 on August, 2 from altitude of 58 m and on October, 12 (68 m). Obtained two orthophotomaps were used for manual vectorization of visible crown projections by means of QGIS and the total area of all projections was calculated. The total area of tree crown projections and the number of trees obtained by GS and UAV methods were matched well. Coniferous trees clearly differed from deciduous trees in aerial photography, also pine was different from spruce. Broadleaf trees (lime and oak) were difficult to detect. For pine the individual crown area obtained by UAV methods was significantly higher and slightly lower for birch and oak. There were no differences for other species. Our results confirmed that UAV technology could be used to obtaining of spatial information on forest characteristics, but tree species identification is still challenging.

Keywords: Unmanned aerial vehicle · Stand mapping · Temperate mixed forests

© Springer Nature Switzerland AG 2019
I. Bychkov and V. Voronin (Eds.): *Information Technologies in the Research of Biodiversity*, SPEES, pp. 48–56, 2019.
https://doi.org/10.1007/978-3-030-11720-7_8

1 Introduction

Over the past decade, the use of Unmanned Aerial Vehicle (UAV) technology in ecosystem studies has proliferated. The spatial resolution of UAV imagery is in the order of a few centimeters per pixel, which allows for observation of fine-scale spatial patterns. UAV surveys are used for a wide variety of environmental applications [1], including vegetation studies [2–4]. In the area of forest ecology UAV technology has the potential to provide a greater amount of spatial information on forest attributes as it allows more rapid surveys of larger areas compared to conventional ground-based surveys. Therefore, UAV-based methods is advantageous for forest monitoring programs and can be complement for traditional forest inventory [5–7]. Data obtained with these methods also can be useful for the development of more accurate simulation models of forest ecosystems dynamics.

For more detailed description of crown competition in the model of growth and elements cycling of temperate forest ecosystems EFIMOD [8] we collected field data from different boreal and mixed forest types on temporary sampling plots by means of tree trunk bases mapping and estimating of the tree height, crown extent and area of crown horizontal projection [9]. Nevertheless, the traditional techniques for measuring crown parameters are very laborious, time-consuming and required appropriate experience. The research reported here is focused on the comparison of a traditional ground-based (GB) method ("from below") and the use of aerial images-based (UAV) method ("from above") by the example of the permanent sample plot in mixed forest located in Moscow region.

2 Materials and Methods

2.1 Study Area

The Prioksko-Terrasny Biosphere Reserve is situated in about 100 km to the south of Moscow, on the left bank of the Oka river (see Fig. 1). The area belongs to the coniferous-broad-leaved forest zone. Total area of the Reserve is about 5000 ha. The Reserve was established in 1945 for protection and study of Moscow region nature, especially for conservation of the "relic Oka-flora". All the Reserve territory was transformed by human activity during last few centuries. The signs of numerous cuttings, plowings, and fires are frequent in the Reserve. Pine (*Pinus sylvestris*), birch (*Betula* spp.), and spruce-broad-leaved forests with *Picea abies, Tilia cordata, Quercus robur* are presented in the most part of the Reserve. There are pine forest plantings in the extensive part. Age of the oldest trees is 160–180 years, age of the main part of the stands is 50–60 years [10].

Our field works were carried out on the permanent sampling plot of 1 ha (100 × 100 m) established in 2016. The plot is aligned from north to south along magnetic meridian. The coordinates of the center of the plot are N 54.88876, E 37.56273 in the WGS 84 datum.

2.2   Ground-Based Methods

All trees with DBH (diameter at breast height) at least of 5 cm were mapped and measured during ground survey in 2016. Mapping was performed with Laser Technology TruPulse 360B angle and distance meter. First, polar coordinates of each tree trunk were measured, and then after conversion to the Cartesian coordinates, the scheme of stand was validated on-site. Species and DBH were determined for each tree. For each alive tree social class (according to Kraft) [11, 12] was detected. Five classes of social status are recognized: (1) dominant, including free-standing trees with upper crown above the general level of the canopy; (2) codominant, which includes trees with crowns forming the general upper level of the canopy; (3) subdominant, which includes trees extending into the canopy and receiving some light from above, but shorter than the dominant and codominant classes; (4) suppressed, including trees with crowns below the general level of the canopy, receiving little direct light from above; (5) dying trees. Also for alive trees the tree height, and radii of crown horizontal projection in four cardinal directions were measured. Crown radii were measured from the outer edge of stem at breast height.

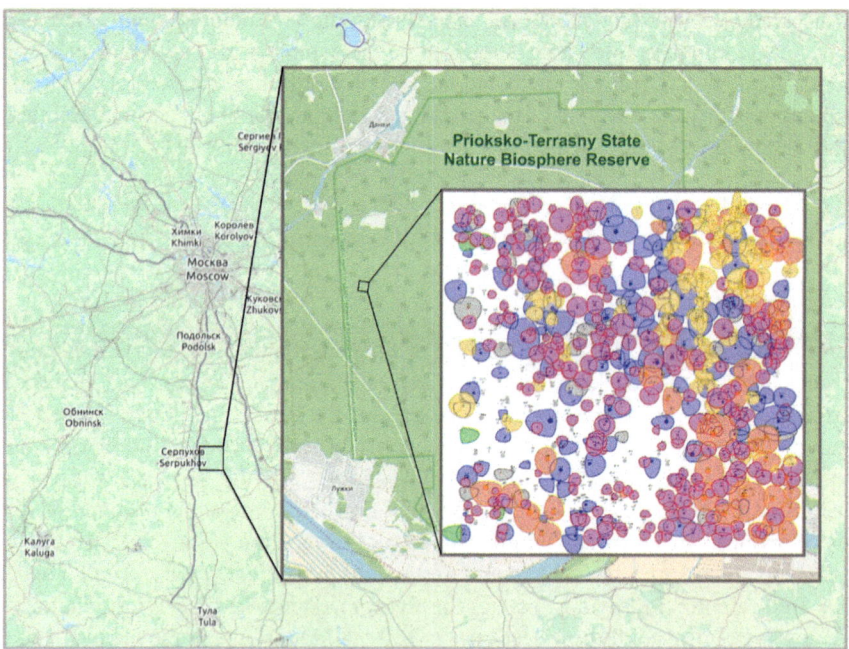

Fig. 1. Location of study area in Moscow region and Prioksko-Terrasny Reserve. Different colors on the scheme of stand mean crowns of different tree species.

2.3   UAV Methods

A lightweight four-wheel UAV Phantom 4 (DJI-Innovations, Shenzhen, China) equipped with an integrated camera of 12Mp sensor was used for aerial photography in this study. Combination of a built-in inertial measurement unit and GPS/GLONASS positioning allow to obtain a much more precise flight position, than using GPS only with accuracy of 2–3 m in the horizontal direction. More detailed parameters are included in Table 1. The aerial photography was conducted in 2017 on August, 2 from altitude of 58 m and on October, 12 (68 m). The commonly used mosaic flight mode was used with 90% overlapping both on side and front directions. The flight plan was drawn with a buffer around the sampling plot for avoiding a processing artifacts, so the final orthophoto image accounted for an area about of 3 ha.

Table 1. The detailed parameters of the camera of DJI Phantom 4.

Parameter	Value	Parameter	Value
Image sensor	1/2.3 in CMOS	RGB color space	sRGB
Camera lens	FOV 94° 20 mm	Shutter speed	1/8000 s
Aperture	f/2.8	Photo size	4000 × 3000 pixels
ISO range	100–1600	Image format	JPEG, RAW

2.4   Sample Data Processing and Analysis

The projection of each tree's crown obtained by ground-based study was represented as a union of 4 quarter-ellipses, and the total area of all projections was calculated and crown were visualized using *R* package *plotrix* [13]. The obtained scheme (see Fig. 1) was georeferenced and vectorized using Quantum GIS 2.8 software [14].

DroneDeploy cloud service [15] was used for aerial photographs processing. This company develop own complicated image stitching algorithm for orthophotomosaics generation. Two sets of aerial photos, 788 images for August 2017 and 501 for October 2017, were processed (see Fig. 2). The obtained orthophotomaps were used for manual vectorization of visible crown projections in Quantum GIS, tree species were identified expertly. The visual interpretation was undertaken by botanist. After that the total area of crown projection polygons and the total area of crown projections of different tree species were calculated.

For comparison we used only alive trees of upper canopy (1, 2, and 3 Kraft's classes), because only the upper canopy is observed on the orthophotomosaics. We compared the total area of tree crown projections and tree count obtained by ground-based tree mapping and UAV methods and also for different species. At the next step we validated the quality of detection of tree locations on the vectorized orthophotomaps. For this we evaluated the percentage of crown overlap between vector layers obtained by two techniques. The tree was considered as correctly detected when the species was matched and the crown overlap was more than half of projection area. For such trees, the paired t-test was used for comparison of individual crown projection area (separately by tree species) obtained by ground survey and UAV methods. The test was performed using R [16].

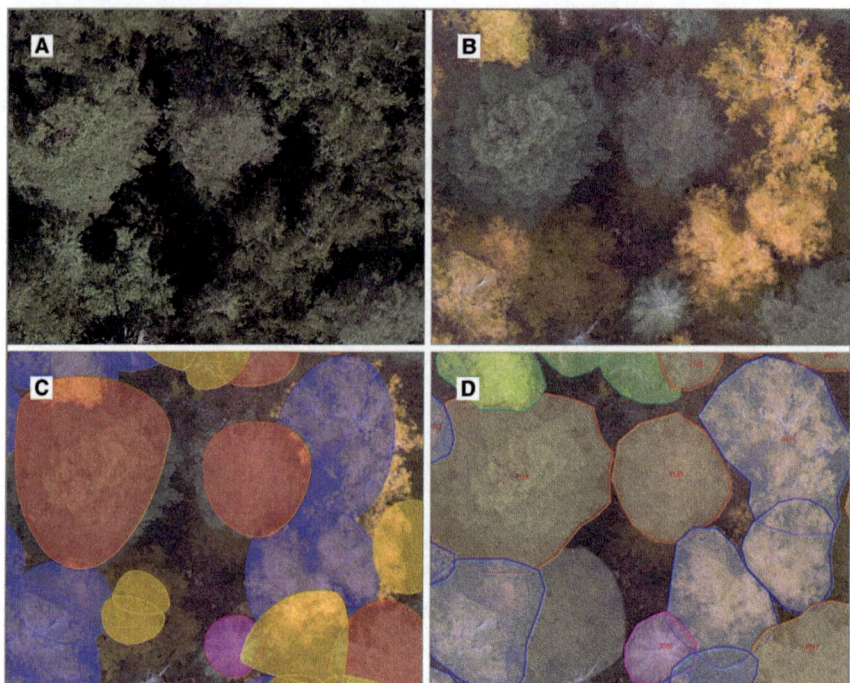

Fig. 2. Fragments of orthomosaic images generated for the two dates (**A**) August, 2 and (**B**) October, 12, and with crown projections visualized using R functions (**C**) and manual vectorization of orthomosaics (**D**).

3   Results

3.1   Permanent Sampling Plot Characteristics

The stand on the permanent sampling plot was mixed. A total of 779 trees, including 582 alive (267 spruces (*Picea abies* (L.) H.Karst.), 47 pines (*Pinus sylvestris* L.), 134 birches (*Betula* spp.), 5 aspens (*Populus tremula* L.), 28 oaks (*Quercus robur* L.), 102 limes (*Tilia cordata* Mill.) and 196 dead individuals (79 spruces, 9 pines, 73 birches, 6 aspens, 27 oaks, 2 limes) were counted during ground survey study. Tree counts of 1–3 Kraft's classes (included in the analysis) are presented in the Table 3. The stand is uneven-aged (see Table 2). Spruce, oak and lime composed regrowth layer. *Calamagrostis arundinacea* Roth, *Pteridium aquilinum* (L.) Kuhn and *Vaccinium myrtillus* L. mainly dominated the ground cover.

3.2   Comparison of the Total Area of Tree Crown Projections

Total area of tree crown projections according to ground-based tree mapping was 6833 m^2 (269 trees of 1–3 Kraft's classes), and according to aerial-based mapping it was 6883 m^2 – about 1% higher (285 vectorized crowns – 6% more).

Table 2. Age characteristics of trees of 1–3 Kraft's classes.

Tree	Minimum age	Maximum age	Mean
Spruce	32	64	78
Pine	69	167	113
Birch	45	165	100
Aspen	80	85	82
Oak	90	134	111
Linden	33	105	65

Among different tree species considered separately the most consistent results among two techniques were obtained for birch and spruce (see Table 3). The total crown area of trees identified as birch during UAVs manual vectorization was 4% less than by ground-based measurements, and the number of trees was 4% more. According to validation results 55% birch trees were recognised correctly. For spruce, the total area of crown projections estimated by aerial photography was 14% less than that estimated by ground-based measurements, whereas the number of trees was 13% higher; 87% spruce trees were identified correctly. For pine total crown area was overestimated by 32%, and the number of trees was 22% higher, while 84% trees were recognised correctly. According to the results of aerial photography, the crown projection area for oak was underestimated by 31%, and the number of trees was less by 13%; 60% trees were recognised correctly. The results for aspen and lime showed high discrepancy between techniques, namely 1.5–2-fold overestimation of both crown projection area and number of trees for aspen by aerial photography, and more than 3-fold underestimation of these parameters for lime. According to validation results one aspen tree only was correctly identified.

Table 3. Comparison of the total area of tree crown projections and tree counts obtained by ground survey (GS) and UAV methods.

	Spruce	Pine	Birch	Aspen	Oak	Lime
Total crown area, m^2						
GS	1365.0	1699.9	3147.3	92.4	308.6	220.1
UAV	1146.0	2237.8	3032.0	195.0	212.5	59.7
Tree counts, ind.						
GS	79	45	112	5	15	13
UAV	89	55	116	8	13	4
Correctly detected	69	38	61	1	9	0

3.3 Comparison of the Area of Crown Projections of Individual Trees

Individual crown projection area of correctly detected pine trees obtained by UAV data were significantly higher (P = 0.0007), than those obtained by ground-based measurements (see Fig. 3). On average the value was overestimated by 12%, but some

crowns were overestimated by 37−60%. There were no significant differences for birch
(P = 0.056) and oak (P = 0.055), but projection areas of individual crowns obtained by
manual vectorization of orthomosaics were slightly lower than calculated based on
ground-survey data. There were no significant differences for spruce (P = 0.667).

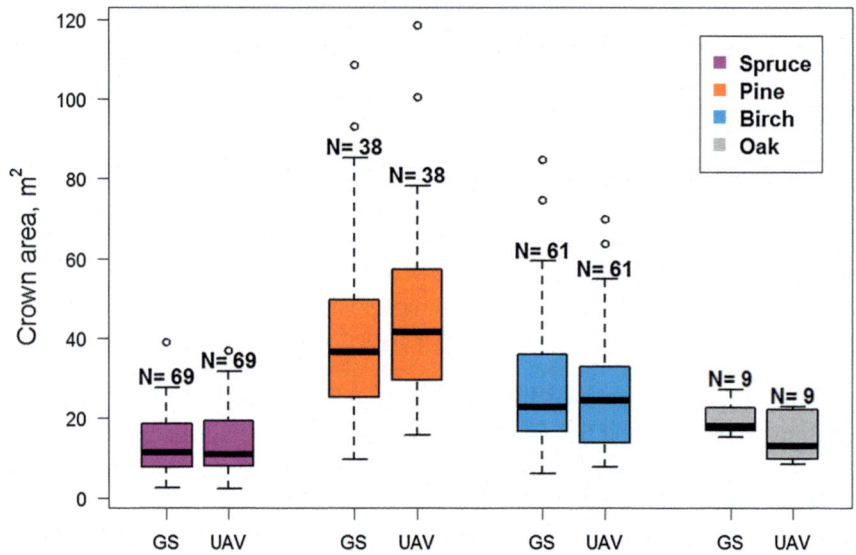

Fig. 3. Comparison of the area of crown projections of individual trees. GS – ground survey
method, UAV – unmanned aerial vehicle method, N – tree counts.

4   Discussion

In this study we demonstrated that lightweight UAVs can be used for assessment of
forest structure characteristics of temperate mixed stands. Our results indicated well-
matched outputs on total area of tree crown projections and the tree counts obtained by
GS and UAV methods.

The quality of species identification and crown projection assessment during the
manual vectorization of orthomosaics depended on tree species. Coniferous trees were
clearly differed from deciduous trees in aerial photography, and also pine was clearly
different from spruce. The total crown area and number of spruce and birch trees
obtained by UAV were in good agreement with ground survey results. Broadleaf trees
(lime and oak) were difficult to detect. The discrepancy in results obtained by two
techniques can be explained by accounting at aerial photography of smaller
(DBH < 5 cm) trees in gaps, which were deliberately omitted during ground survey.
One more reason can be the uncertainty of recognition of similarly looking aspen, birch
and lime on aerial photo.

The under- or overestimation of crown projection area with aerial photography
could be a result of not exactly correct rough approximation of crown projection shape

during ground survey. This assumption is confirmed by the results of paired comparisons of individual crown projections (for correctly detected trees) obtained by two methods. For pine the area of crown projection obtained from UAVs data was significantly higher than those obtained by ground survey. Probably, the projection of the crown drawn on the basis of four radii does not give a correct estimation of its area. Moreover, the parts of single tree's crown, branched at small height above the ground, could be recognized on orthophotomaps as individual crowns.

For oak and birch the trend was opposite. Crown projection areas of individual trees obtained by UAV were less than those obtained by GS data. Probably, birch crowns have more overlapping with neighbor trees compared to other tree species. As we can see only the upper canopy is observed on the orthomosaics, and overlaps were not taken into account. All oak trees were extending into the canopy (3 Kraft's class), their crowns were located below the general level of the canopy and therefore were poorly visible on the orthomosaic images.

Acknowledgements. The   work   was   supported   by   Russian   Science   Foundation (project No. 18-14-00362).

References

1. James, M.R., Robson, S., d'Oleire-Oltmanns, S., Niethammer, U.: Optimising UAV topographic surveys processed with structure-from-motion: ground control quality, quantity and bundle adjustment. Geomorphology **280**, 51–66 (2017). https://doi.org/10.1016/j.geomorph.2016.11.021
2. van Iersel, W., Straatsma, M., Addink, E., Middelkoop, H.: Monitoring height and greenness of non-woody floodplain vegetation with UAV time series. ISPRS J. Photogrammetry and Remote Sens. **141**, 112–123 (2018). https://doi.org/10.1016/j.isprsjprs.2018.04.011
3. Rossini, M., Di Mauro, B., Garzonio R., Baccolo, G., Cavallini, G., Mattavelli., M., Colombo R.: Rapid melting dynamics of an alpine glacier with repeated UAV photogrammetry. Geomorphology **304**, 159–172 (2018). https://doi.org/10.1016/j.geomorph.2017.12.039
4. Zhang, H., Sun, Y., Chang, Li., Qin, Y., Chen, J., Qin, Y., Du, J., Yi, S., Wang, Y.: Estimation of grassland canopy height and aboveground biomass at the quadrat scale using unmanned aerial vehicle. Remote sensing **10**(6), 851 (2018). https://doi.org/10.3390/rs10060851
5. Otero, V., Van De Kerchove, R., Satyanarayana, B., Martnez-Espinosa, C., Fisol, M.A.B., Ibrahim, M.R.B., Sulong, I., Mohd-Lokman, H., Lucas, R., Dahdouh-Guebas, F.: Managing mangrove forests from the sky: Forest inventory using field data and Unmanned Aerial Vehicle (UAV) imagery in the Matang Mangrove Forest Reserve, peninsular Malaysia. For. Ecol. Manage **411**, 35–45 (2018). https://doi.org/10.1016/j.foreco.2017.12.049
6. Zahavi, R.A., Dandois, J.P., Holl, K.D., Nadwodny, D., Reid, J.L., Ellis, E.C.: Using lightweight unmanned aerial vehicles to monitor tropical forest recovery. Biol. Conserv. **186**, 287–295 (2015). https://doi.org/10.1016/j.biocon.2015.03.031
7. Zhang, J., Hu, J., Lian, J., Fan, Z., Ouyang, X., Ye, W.: Seeing the forest from drones: testing the potential of lightweight drones as a tool for long-term forest monitoring. Biol. Conserv. **198**, 60–69 (2016). https://doi.org/10.1016/j.biocon.2016.03.027

8. Komarov, A.S., Chertov, O.G., Zudin, S.L., Nadporozhskaya, M.A., Mikhailov, A.V., Bykhovets, S.S., Zudina, E.V., Zoubkova, E.: EFIMOD 2–a model of growth and elements cycling of boreal forest ecosystems. Ecol. Model. **170**, 373–392 (2003)
9. Shanin, V.N., Shashkov, M.P., Ivanova, N.V., Grabarnik, P.Y.: The effect of aboveground competition on spatial structure and crown shape of the dominating canopy species of forest stands of European Russia. Russian J. Ecosyst. Ecol. **1**(4) (2016). https://doi.org/10.21685/2500-0578-2016-4-5. (in Russian)
10. Smirnova, O.V., Shaposhnikov, E.S. (eds.): Forest successions in protected areas of Russia and problems of biodiversity conservation. Russian Bot. Soc., St-Petersburg (1999). (in Russian)
11. Lakatos, F., Mirtchev, S.: Manual for visual assessment of forest crown condition. Food and Agriculture Organization of the United Nations, Pristina (2014). http://www.fao.org/3/a-i4214e.pdf
12. Eichhorn, J., Roskams, P., Potočić, N., Timmermann, V., Ferretti, M., Mues, V., Szepesi, A., Durrant, D., Seletković, I., Schröck, H-W., Nevalainen, S., Bussotti, F., Garcia, P., Wulff, S.: Manual on methods and criteria for harmonized sampling, assessment, monitoring and analysis of the effects of air pollution on forests. Thünen Institute of Forest Ecosystems, Eberswalde, Germany (2016). https://www.icp-forests.org/pdf/manual/2016/ICP_Manual_2017_02_part04.pdf
13. Lemon, J.: Plotrix: a package in the red light district of R. R-News. **6**(4), 8–12 (2006)
14. QGIS Development Team. QGIS Geographic Information System. Open Source Geospatial Foundation Project, http://qgis.osgeo.org. Last accessed 31 Aug 2018
15. DroneDeploy, https://www.dronedeploy.com/. Last accessed 31 Aug 2018
16. R Core Team. R: A language and environment for statistical computing. R Foundation for Statistical Computing, Vienna, Austria, URL https://www.R-project.org/. Last accessed 31 Aug 2018

Creation of Information Retrieval System on the Unique Research Collections of the Zoological Institute RAS

Oleg Pugachev, Natalia Ananjeva, Sergey Sinev, Leonid Voyta,
Roman Khalikov⑩, Andrey Lobanov, and Igor Smirnov^(✉) ⑩

Zoological Institute of the Russian Academy of Sciences, 199034 Saint
Petersburg, Russia
{pugachev,natalia.ananjeva,sinev,leonid.voyta,roman.
khalikov,all,smiris}@zin.ru

Abstract. Zoological Institute of the Russian Academy of Sciences (ZIN)—is one of the oldest scientific institutions in Russia with unique scientific collections. In 2017, the Museum and the Institute celebrated their 185th anniversary. The unique collection of the ZIN were collected by many generations of Russian and foreign zoologists throughout the territories and waters of our planet. Collected during these expeditions materials are stored in the Institute, they remain imperishable source of information on the structure and distribution of faunal diversity in space and time. The collections continue to rise. ZIN has one of the largest zoological collections in the world, with more than 60 million of storage units. In general, in the collections of the ZIN about 260 thousand species of animals, which is about a quarter of the known world fauna, are presented. There is almost all the animal species inhabiting territory and waters of Russia, for many of them large series are stored. Several tens of thousands of type specimens of animal species stored in the collection are of exceptional value. The content of the concept of "zoological collection" in our days of rapid progress of information technologies and advances in molecular genetic studies quickly filled with new meanings. Currently, the collection is rightfully regarded as a bank of scientific information and the primary tool for basic and applied biological research. The information system should include taxonomical, type, collection, zoogeographical and bibliographic data. By using of available server infrastructure of ZIN and information system of collection specimens created (ZIN Research Collections portal, https://www.zin.ru/collections/), there was a possibility of selective publication of ZIN research collection data via GBIF portal (http://ipt.zin.ru). The scientific use of the collections and depositories of integrated information systems for faunistic biodiversity in the present context includes, as the most important modern goals, digitization of collections and publication of information in the public domain on the Internet.

Keywords: Zoological collections · Zoological Institute of the Russian Academy of Sciences · Studying and saving of biological diversity · Information systems · Virtual (digital) museum · Digitalization

© Springer Nature Switzerland AG 2019
I. Bychkov and V. Voronin (Eds.): *Information Technologies in the Research of Biodiversity,* SPEES, pp. 57–65, 2019.
https://doi.org/10.1007/978-3-030-11720-7_9

1   Research Collections of the Zoological Institute RAS

Zoological Institute of RAS—is one of the oldest scientific institutions in Russia with unique scientific collections. Their history is connected with the organization of the first Russian museum—Kunstkamera by Peter I in 1714. From the Kunstkamera in 1832 Zoological Museum of the Imperial Academy of Sciences was separated as a distinct institution which in the XX century was reorganized into the Zoological Institute of the USSR Academy of Sciences (since 1991—Russian Academy of Sciences) by the Decree of the General Meeting of the Academy of Sciences of the USSR on December 26, 1931.

In 2017, the Museum and the Institute celebrated their 185th anniversary [1]. The unique collection of the Zoological Institute were collected by many generations of Russian and foreign zoologists.

Materials collected during these expeditions are stored in the Zoological Institute of the Russian Academy of Sciences; they remain imperishable source of information on the structure and distribution of faunal diversity in space and time. Zoological collections continue to rise, primarily due to samplings during numerous expeditions of the Institute, as well as the contributions from a variety of companies and institutions (Russian Geographical Society, the Society for the Study of Siberia, the Office of the Northern Sea Route, the various nature reserves and others), by zoologists, local naturalists, nature lovers, through the exchange with other museums, donations from individuals and in special cases by purchase.

Zoological Institute has one of the largest zoological collections in the world, with more than 60 million of storage units. Storage units—is any object that has a label with information about the place and time of collection, collector's name and ideally—with the scientific identification. Storage units are stuffed carcasses, skins, skeletons of vertebrate animals and their parts, birds' eggs and nests, dry and wet (in alcohol or formalin), fish, amphibians, reptiles and invertebrates, special preparations of animals of microscopic size or their parts, modern and fossils remains of animals, DNA samples or individual sequencing. The concept of «storage unit» has many equivalents, some of which (samples of zoological collections, scientific materials, specimens and so on) are used in the juridical and normative vocabulary.

The zoological collection is an ordered set of documented scientific objects representing the scientific or educational interest. The structure of the collections includes review and research collections. Review exposition of different types of animals, integrated into Zoological Museum, has about 30,000 specimens and is one of the world's largest [2].

Research collections are presented by the systematic collection and monitoring collections (repetitive sampling in order to identify changes in natural communities). They are stored in the leading scientific centers of the world—such as the Museum of Natural History in London (UK), the National Museum of Natural History in Paris (France), the National Museum of Natural History in Washington, DC (Smithsonian Institution) and the American Institute of Natural History, New York (USA), Senckenberg Museum in Frankfurt am Main (Germany), Museum of Natural History in Leiden (Netherlands) and many other institutions and museums. Zoological Institute in

St. Petersburg is one of the first institutions among the world's depositories of collections of animals and is the largest in Russia and CIS both in number of specimens and representation of faunal diversity [2].

In general, collections of the Zoological Institute hold about 260 thousand species of animals, which is about a quarter of the known world fauna. There are almost all the animal species inhabiting territory and waters of Russia, for many of them large series are stored.

For many groups of animals from the Northern Hemisphere of the Old World this is the most representative collection in the world. Unique collection of Zoological Institute is included in the global network of zoological collections as an integral part of the actual scientific basis for zoologists around the world. Every year specialists from dozens of countries visited Zoological Institute.

Research of the staff of the Zoological Institute, including the sampling of collections, allowed today substantially revise the outlook on biodiversity in many regions of the world, for example such world centers of biodiversity as South East Asia and the Southern Ocean [3].

Several tens of thousands of type specimens of animal species stored in the collection are of exceptional value. Type specimens have the status of international standards and are the objective basis for Zoological Nomenclature. These type specimens (standards) animals in its uniqueness and significance can be compared only with the standards of weights and measures. Type specimens are unique by definition, i.e., have no analogues and cannot be replaced.

It is important to note that the expansion of the technical capabilities of the methods of DNA extraction (including the work with so-called «ancient» DNA) already allows in the near future will significantly increase the scope of historically significant collection specimens, including type specimens and subfossil samples that will have the critical value for many aspects of biodiversity research.

The content of the concept of "zoological collection" in our days of rapid progress of information technologies and advances in molecular genetic studies quickly filled with new meanings. Currently, the collection is rightfully regarded as a bank of scientific information and the primary tool for basic and applied biological research.

2 Computerization of the Zoological Institute

More than 40 years have passed since zoologists of the Kaliningrad biological station have started to cooperate with the group of scientists of the Ioffe Physical-Technical Institute of the Russian Academy of Science, to create a database (DB) on birds ringing and to develop the software for support and operation of this database [4, 5]. In 1977 under A.F. Alimov's initiative physicist A.A. Umnov has been invited to the Institute. BESM-6 modeling calculations of productional processes have begun. In 1982 owing to A.A. Umnov's diligence, a terminal of BESM-6 has appeared in ZIN. In 1985, again on the initiative of A.F. Alimov, A.L. Lobanov was invited. He initiated the development of zoological DB and machine identification keys.

In 1988, the Institute receives the electronic computer SM-1420 that allows expanding the circle of employees interested in using computers in zoology.

Finally, the purchase of personal computers (the first Amstrad have appeared in 1989) became the beginning of wide introduction of computer methods in zoological researches. In 1990 ZIN held a meeting on DB in biology [5].

In 2017, Russian Fund of Basic Researches (RFBR) celebrated its 25th anniversary. In 1993 the first grant of the Russian Federal Property Fund 93-04-21216 with computerization support has been received (System of the computer integrated data processing on animals biodiversity (ZOOINT—ZOOlogical Integrated Information Retrieval System). "ZOOINT" is a long-lived project, which used standard ZOOCOD as taxonomic basis (the specialized algorithm for representing hierarchical data in flat tables of relational DB), served as the sample for a lot of other projects [5–7, 9, 10].

During the years of computerization, various projects based on modern information technologies have been implemented at the Zoological Institute. Creation of the Internet-portal of the Institute, an intelligence-analytical system on fleas of the Globe and base and databanks on various parasites, the Information Retrieval System (IRS) "OCEAN", entomological DB, dialogue computer diagnostic systems, the IRS for fresh-water fishes of Russia, virtual collections («Protists») and museums on the Internet, electronic publications and technology of hyperbases of data, usage of geographical intelligence systems (GIS) for the analysis of areas, the IRS "Biodiversity of Russia", the IRS "Biodiversity of animals of Russia", DB of the White Sea biological station, mathematical and imitating models in production biology are only a small part of them [1, 5, 11, 12].

The scientific use of depositories of collections and integrated information systems for faunistic biodiversity in the contemporary context includes, as the most important modern goals, digitization of collections and factual data placement in the public domain on the Internet as well as creation of a DNA bank. The all this will allow collecting complete information on the existing biodiversity, where each component part—collections, DNA bank and DB of specimens—will carry their own specific functions and can adequately serve the needs of biological science and technology in the new millennium.

3 Digitized Research Collections of the Zoological Institute RAS

The study of the biological diversity of animals in all its aspects is a fundamental scientific task, solved within the framework of an interdisciplinary complex of systematic, zoogeographical, ecological, molecular-genetic studies and the increasing application of information technologies every year. Research zoological collections are the most important tool and information basis for this kind of research. At present, these unique materials require active involvement in scientific circulation through the creation of integrated information systems on biodiversity, developed by zoologists in the process of scientific supervision of collections by their main groups of fauna.

One of the most urgent scientific problems, as well as a necessary condition for the modernization of zoological collections and algorithms for their research, is the development of information systems on biodiversity and research collections with subsequent integration into international distributed information retrieval systems [9, 13–16].

Now the Zoological Institute of the Russian Academy of Sciences has started realization of a specific fundamental task, which consists in developing an algorithm for digitizing research collections of the Zoological Institute, considering the specificity of their storage for separate systematic groups. In 2015, work began on the digitization of ZIN research collections within the framework of the RFBR project 15-29-02457 "The collections of the Zoological Institute as an important tool and information basis of fundamental biological research" [14].

Considering the huge amount of this collection (more than 60 million units of storage and tens of thousands of copies of primary types), the first stage of work is limited to a number of model taxa belonging to the main classes of the animal world. Priority attention is given to the reference (voucher) specimens, which are carriers the scientific names of species and subspecies. The search for such specimens, their correct designation and research using the most up-to-date methods is a scientific task of primary importance, since any inaccuracies made in the typification of taxa can lead to a chain of erroneous conclusions at all levels of biological diversity research and substantially distort, or even completely devaluate the results. It is particularly relevant for typical specimens of species established in the 18th and 19th centuries even before the creation of the International Code of Zoological Nomenclature, often not fully described and specifically not identified, which creates serious problems in identifying taxa and the proper application of their names.

The following groups of terrestrial and aquatic invertebrates were used as model groups for development of the catalog collection structure and algorithms: Pogonophora, Coleoptera, Lepidoptera, Siphonaptera, Hymenoptera, Diptera, Simuliidae, Ceratopogonidae, Asteroidea, Ophiuroidea, Echinoidea, Holothuroidea, Amphibia, Reptilia, and Mammalia.

The information system is built on a modular architecture and includes taxonomic, type, collection, zoogeographic and bibliographic information. Of great importance is the module of high-quality graphic images library of specimens and related materials. Key steps in performing the work include digitalization of ZIN collection catalog records, Latinization of specimen's localities, georeferencing of specimen's localities, obtaining high-quality images of storage objects. Particular importance is given to the preservation of both information in the authentic form (data as they are in the primary source) and information on processing (interpretation) proposed by different experts and different methods.

Digitization gets special importance for type specimens, multiplying their availability for the scientific community many times over. The information system created within the framework of the project initially provides presence of library of collection sample images and additional materials. To obtain high-quality images varieties of digitalization methods are used. These include scanning of collection specimens and related materials (original labels, catalog cards, storage units, envelopes, x-ray pictures, etc.) on flatbed CCD-scanners of the high resolution, as well as specialized photography (macrophotography) using digital SLR cameras in the kit with special macro-lenses and systems of pulsed light sources (macro flashes) [17]. As a result, it allows using the single unified storehouse of high-quality images of the high and ultrahigh resolution as archival data storage to work with them in the intranet and for online data publishing.

The created information retrieval system is based on advanced domestic and international standards. The kernel of all IT zoological projects at the Zoological Institute is ZOOCOD (the data standard for taxonomic table's creation and representation of multilevel hierarchies in relation databases) with ZooDiv taxonomic classifier (146653 taxonomic records of 40 ranks) [18]. International standards for georeferencing of findings points by National Science Foundation, application-oriented interfaces of online mapping Google Maps API are also used, making an opportunity of the subsequent integration of the created information system into distributed zoological online resources. The infological structure of the information system is constructed on the client-server architecture, includes server database management system (MS SQL Server—server side) and universal web-interface (client side—ASP, cross-platform script language JavaScript, AJAX, Open Graph Protocol, etc.).

Digitized research collections are published online for public access. For this purpose, a specialized web site was established on the ZIN RAS web portal [19]. The website was initially implemented in two language versions—Russian and English. The English version completely duplicates the Russian version, both in terms of data presenting and in terms of functionality. The full bilingualism of the site significantly increases the relevance of this resource to foreign colleagues, making the availability of research collections information, including the most important data on type specimens, to a qualitatively new level. The site is equipped with a simple but functional search and optimized in accordance with the standards of the Open Graph Protocol to ensure the full integration of the unique content of the site with various commonly used and specialized social networks and its correct citation in them.

The functionality of the website expanded by the mechanism for exporting sample collection specimens. This mechanism uses international standard Darwin Core and the full data on collection specimens can be downloaded as XML documents. The mechanism for exporting collection data is completely bilingual and unified for all model collection groups.

The existing server infrastructure and the created information system allowed deploying a special service—Integrated Publishing Toolkit, IPT [20]. This enables the selective publication of ZIN collection data to the portal of Global Biodiversity Information Facility (GBIF). To date, five datasets for the type specimens of different groups have been published: Ophiuroidea [21], Polycestinae [22], Cosmopterigidae [23], Pogonophora [24], Bufonidae and Megophryidae [25]. Collections data integration between the Zoological Institute and GBIF will continue.

Acknowledgements. Work is performed within the framework of the following projects and state themes: AAAA-A17-117030310207-3, AAAA-A17-117030310017-8 and AAAA-A17-117080110040-3; also it was partially supported by grants of the Russian Foundation for Basic Research № 15-04-02971 and № 15-29-02457, and the program of Presidium of the Russian Academy of Sciences on studying a biological diversity "Taxonomical researches of animals as a basis of inventory of a biodiversity and its information supply".

References

1. Alimov, A.F., Ananjeva, N.B., Dianov, M.B., Lobanov, A.L., Pugachev, O.N., Ryss, A.Y., Smirnov, I.S., Khalikov, R.G.: The role of the information technologies in the researches of the Zoological Institute of the RAS. Materials of the Jubilee Reporting Scientific Session on the 185th Anniversary of the Zoological Institute of the Russian Academy of Sciences (St. Petersburg, Russia, November 13–16, 2017). Collected articles. Zoological Institute RAS. SPb., 2017: 5–8. (In Russian)
2. Alimov, A.F., Tanasijchuk, V.N., Stepanjants, S.D.: A diversity of the World fauna in collections of the Zoological Institute of the Russian Academy of Science. In: Petrosjan, J. A., Ivanova, E.A. (eds.) The Treasures of the Russian Academy of Sciences Collections in St. Petersburg. SPb. Science pp. 239–298 (2003). (In Russian)
3. Neyelov, A.V., Sirenko, B.I., Smirnov, I.S., Gagaev, S.Y., Golikov, A.A., Andreev, M.P., Gavrilo, M.V.: To the 60th anniversary of the Russian studies of Antarctic ecosystems. Materials of the Jubilee Reporting Scientific Session on the 185th Anniversary of the Zoological Institute of the Russian Academy of Sciences (St. Petersburg, Russia, November 13–16, 2017). Collection of articles. Zoological Institute, RAS. SPb., pp. 130–134 (2017). (In Russian)
4. Morozov, J.G., Efremov, V.D.: The software for support and operation of ornithological databases. Databases and the Computer Graphics in Zoological Researches (Works of Zoological institute, т. 269), pp. 91–99 (1997)
5. Lobanov, A.L., Smirnov, I.S.: Place and role of the information technologies in the Zoological Institute researches. Fundamental Zoological Researches: Theory and Methods, pp. 283–318. KMK Scientific Press Ltd., Moscow-St.-Petersburg 2004. (In Russian with English summary)
6. Lobanov, A., Sokolov, E., Smirnov, I.: ZOOINT—an integrated system for zoological data bases. Proceedings of the International Workshop on Advances in Databases and Information Systems, pp. 270, 271. Moscow (1994) 23–26 May 1994
7. Smirnov, I.S., Lobanov, A.L., Alimov, A.F., Medvedev, S.G., Golikov, A.A.: The results of development of the project ZOOINT and its future perspectives. Digital Libraries: Advanced Methods and Technologies, Digital Collections: Proceedings of the Fourth All-Russian Scientific Conference RCDL'2002 (Dubna, October 15–17, 2002), pp. 308–315. In JINR,– V.2, 2 v., Dubna (2002). (In Russian with English summary)
8. Lobanov, A.L., Smirnov, I.S., Dianov, M.B., Golikov, A.A., Khalikov, R.G.: Evolution of the standard ZOOCOD–concept of reflection of zoological hierarchical classifications in the plane tables of relation databases. Digital Libraries: Advanced Methods and Technologies, Digital Collections: Proceedings of the Tenth All-Russian Scientific Conference RCDL'2008 (Dubna, October 7–11, 2008), pp. 326–332. JINR, Dubna (2008). (In Russian with English summary)
9. Lobanov, A.L., Kireitchuk, A.G., Pugachev, O.N., Smirnov, I.S.: Zoological collections, interactive identification keys and the Internet. Zoological Collections of Russia in XVIII–XXI Centuries: A Sociopolitical and Scientific Context. Editor-compiler N.V.Slepkova. SPb: Publishing House SPbGETU "LETI", pp. 123–132 (2012). (In Russian with English summary)
10. Smirnov, I.S., Lobanov, A.L., Pugachev, O.N., Kireitchuk, A.G., Krivokhatskiy, V.A., Dianov, M.B., Khalikov, R.G., Golikov, A.A., Voronina, E.P.: From system ZOOINT to system ZOODIV. Digital Libraries: Advanced Methods and Technologies, Digital Collections: Proceedings of the XIV All-Russian National Research Conference

RCDL'2012. Pereslavl-Zalesskij, Russia, 15–18 Oct 2012, pp. 241–246. Pereslavl University, Pereslavl-Zalesskij (2012). (In Russian with English summary)

11. Pugachev, O.N., Alimov, A.F., Lobanov, A.L., Krivokhatskiy, V.A., Smirnov, I.S.: The first results of the development of the information system on biodiversity of Russia (BioDiv-ZooDiv). Information systems and web-portals on diversity of species and ecosystems. Proceedings of the International Symposium, Borok, on November 28–December 1, pp. 170–173. KMK Scientific Press LTD, Moscow (2006). (In Russian)

12. Krivokhatsky, V.A., Lobanov, A.L., Medvedev, G.S., Belokobylsky, S.A., Dianov, M.B., Smirnov, I.S., Khalikov, R.G.: Information system on the entomological collection in Internet. Proceedings of Russian Entomological Society, vol. 74, pp. 59–70 (2003). (In Russian)

13. Khalikov, R.G.: ZooDiv–an information storage and retrieval biodiversity system and its implementation for varied zoological projects. The Russian-Chinese Seminar «Research and Protection of Amphibians and Reptiles of Eurasia: Results and Prospects of Cooperation», on July, 29th–on August, 3rd, 2009, p. 19 St.-Petersburg, (2009)

14. Ananjeva, N.B., Voyta, L.L., Volkovitsh, M.G., Golikov, A.A., Dianov, M.B., Medvedev, S.G., Petrova, E.A., Sinev, S.Y., Smirnov, I.S., Smirnov, R.V., Syromjatnikova, E.V., Khalikov, R.G., Khalin, A.V., Shumeev, A.N.: Research Collections of the Zoological Institute RAS as Resource of Biodiversity Study. Accounting of scientific session on results of works 2015. Abstracts. 12–14th April 2016, pp. 9–10. Zoological Institute (2016). (In Russian)

15. Smirnov, I.S., Lobanov, A.L., Pugachev, O.N., Alimov, A.F., Voronina, E.P.: Digital Collection in Zoology and Digital Libraries. Digital libraries, vol. 9(4), pp. 115–118 (2006). (in Russian). http://www.elbib.ru/index.phtml?page=elbib/rus/journal/2006/part4/SLPAV

16. Smirnov, I.S., Pugachev, O.N., Kireitchuk, A.G., Dianov, M.B., Lobanov, A.L., Khalikov, R.G., Golikov, A.A., Krivokhatskiy, V.A.: Results and prospects of development of information system on a biodiversity of animals of Russia (ZOODIV–BIODIV). Digital Libraries: Advanced Methods and Technologies, Digital Collections: Proceedings of the Twelfths All-Russian Scientific Conference RCDL'2010 (Kazan, October 13–17, 2010), pp. 461–464. Kazan University, Kazan (2010). (In Russian with English summary)

17. Khalikov, R.G., Orlov, N.L.: Experience of practical use of modern digital imaging techniques in zoological research. Annual Reports of ZIN RAS, p. 47–50 (2001)

18. Taxonomic Classifier of Animals, http://www.zin.ru/ZooDiv/animals.asp. Last accessed 14 Aug 2018

19. The ZIN RAS Web Portal on Collections, https://www.zin.ru/collections/index_en.html. Last accessed 14 Aug 2018

20. Integrated Publishing Toolkit, IPT in the Zoological Institute, http://ipt.zin.ru. Last accessed 14 Aug 2018

21. Smirnov, I.S., Golikov, A.A., Khalikov, R.G.: Ophiuroidea collections of the Zoological Institute Russian Academy of Sciences. Zoological Institute, Russian Academy of Sciences, St. Petersburg. Dataset/Checklist. http://ipt.zin.ru:8080/ipt/resource?r=zin_ophiuridae; https://doi.org/10.15468/ej3i4f (2015). Last accessed 14 Aug 2018

22. Volkovitsh, M.G., Golikov, A.A., Khalikov, R.G.: Catalogue of the type specimens of Polycestinae (Coleoptera: Buprestidae) from research collections of the Zoological Institute, Russian Academy of Sciences. Zoological Institute, Russian Academy of Sciences, St. Petersburg. Dataset/Checklist. http://ipt.zin.ru:8080/ipt/resource?r=zin_polycestinae. https://doi.org/10.15468/c3eork (2016). Last accessed 14 Aug 2018

23. Sinev, S.Y., Golikov, A.A., Khalikov, R.G.: Catalogue of the type specimens of Cosmopterigidae (Lepidoptera: Gelechioidea) from research collections of the Zoological Institute, Russian Academy of Sciences. Zoological Institute, Russian Academy of Sciences,

St. Petersburg. Dataset/Checklist. http://ipt.zin.ru:8080/ipt/resource?r=zin_cosmopterigidae. https://doi.org/10.15468/sbga6b (2016). Last accessed 14 Aug 2018

24. Smirnov, R.V., Golikov, A.A., Khalikov, R.G.: Catalogue of the type specimens of Pogonophora (Annelida; seu Polychaeta: Siboglinidae) from research collections of the Zoological Institute, Russian Academy of Sciences. Zoological Institute, Russian Academy of Sciences, St. Petersburg. Dataset/Checklist. http://ipt.zin.ru:8080/ipt/resource?r=zin_ pogonophora. https://doi.org/10.15468/1mlkdp (2017). Last accessed 14 Aug 2018

25. Milto K.D., Ananjeva N.B., Golikov A.A., Khalikov R.G.: Catalogue of the type specimens of Bufonidae and Megophryidae (Amphibia: Anura) from research collections of the Zoological Institute, Russian Academy of Sciences. Zoological Institute, Russian Academy of Sciences, St. Petersburg. Dataset/Checklist. http://ipt.zin.ru:8080/ipt/resource?r=zin_ megophryidae_bufonidae. https://doi.org/10.15468/crgfcq (2016). Last accessed 14 Aug 2018

The Use of Mathematical Methods in Analysis of Antibioticresistans of Microorganisms of Lake Baikal

E. V. Verkhozina[1], A. S. Safarov[2(✉)], V. A. Verkhozina[3],
and U. S. Bukin[4]

[1] Institute of the Earth's Crust Siberian Branch of the Russian Academy of
Science, Irkutsk, Russia
[2] L. A. Melentiev Energy Systems Institute of Siberian Branch of the Russian
Academy of Science, Irkutsk, Russia
alexssss@list.ru
[3] National Research Irkutsk State Technical University, Irkutsk, Russia
[4] Limnological Institute Siberian of Branch of the Russian Academy of Science,
Irkutsk, Russia

Abstract. It is shown that the analysis of a large number of the obtained data, it
is possible variance and correlation methods. It allows to reveal fluctuations of
antibiotic resistance in seasonal and interannual differences also calculate pairwise correlation coefficients. It is identified that the correlation between the
number of bacteria and antibiotic resistance is absent.

Keywords: Dispersion method · Correlation analysis · Antimicrobial
resistance · Aquatic organisms · Ecosystem · Lake Baikal · First section

1 Introduction

One of the most pressing problems in environmental studies is the detection of
anthropogenic factor in the pollution of natural fresh water. Lake Baikal is a complex
ecosystem consisting of several subsystems. Especially the subsystems of the pelagic
and littoral of the lake differ because of the huge difference in depths and temperature
factor. These parameters determine the development of biota, the natural factor and the
self-cleaning ability of ecosystems to anthropogenic factor. Because of the complexity
of turbulent currents, the phenomenon of "patching" in the Baikal ecosystem [1]. It is
very difficult to identify the nature of the supply of organic matter to the Baikal
ecosystem. The search for indicators of anthropogenic impact on aquatic ecosystems is
a very difficult task.

At present, the relevance of microbiological monitoring of water bodies is
becoming one of the most urgent problems. Many researchers found that one of the
most important biological characteristics of microorganisms is their resistance to
antimicrobial drugs [2–4]. The study of microbial communities of water bodies and the
determination of antibiotic resistance of bacteria showed that in reservoirs where
anthropogenic influence is observed, favorable conditions are created for the formation

I. Bychkov and V. Voronin (Eds.): *Information Technologies
in the Research of Biodiversity,* SPEES, pp. 66–72, 2019.
https://doi.org/10.1007/978-3-030-11720-7_10

and preservation of resistant strains of microorganisms. In addition, it is important that under the influence of human activities there are significant violations of evolutionary aquatic biocenoses. There is an increase in conditionally pathogenic microorganisms, their structural changes, their species diversity is significantly changing. In addition, there is a substitution of antibiotic-sensitive microflora for resistant [5, 6]. It should also be noted that, from various environmental objects, water and, consequently, the waterway of transmission of infection are a particular epidemiological danger. To assess water quality by prevalence and species diversity of opportunistic microorganisms, the level of their antibiotic resistance is [7]. The prevailing opinion among some epidemiologists that the external environment is a cemetery for pathogenic bacteria is currently undergoing significant changes.

2 Materials and Methods

Sampling of water was carried out in the littoral zone of the ecosystem of Lake Baikal (Listvyanka, Baikalsk, Slyudyanka) from 2001. In different biological seasons. The sampling procedure was carried out in accordance with GOST 51592–2000. The treatment of the material and the determination of the total microbial number (MFD) corresponded to the generally accepted approaches and methods [8]. During the research more than 3500 strains of bacteria were isolated and analyzed. Identification of isolated microorganisms and determination of their resistance to 12 antimicrobial preparations of six pharmacological groups: penicillins, cephalosporins, aminoglycosides, fluoroquinolones, tetracyclines, diaminopyrimidines, were performed by dilution in MPA (meat-peptone agar) in accordance with the generally accepted methods of the State Sanitary Epidemiological Service of the Russian Ministry of Health [9]. In processing the results obtained, the correlation analysis method [10–12] was applied using standard parametric and nonparametric criteria, as well as a package of computer programs "Statistica". To determine the cause-effect relationships, the obtained data were processed using the dispersion and correlation methods of analysis. This method allows one to determine the strength and direction of variability between variables in a vast array of obtained data as a result of the studies, and is considered one of the promising [13, 14]. For statistical processing and data visualization, a freely distributed programming environment R was used. The evaluation of the data sampled in the work for compliance with the law of normal distribution was carried out using the Shapiro-Wheelk test [15]. During the statistical analysis, the entire array of data was averaged, for each sampling point at a certain point in time, the average resistance of the community to antibiotics was assessed. The data set was grouped separately for each factor: the sampling month and the year of sampling. For each of the factors, a one-way analysis of variance was carried out using the Kraskel-Wallis method. H0 hypothesis for the criterion - the average value of the indicator in the samples is not significantly different (the factor in question does not affect the average indices of the samples). H0 deviated with the estimated probability of its acceptance $P\_value < \alpha = 0.05$. The result was visualized using bar charts. For each antibiotic, the coefficients of variation in the resistance of bacterial strains to it from the investigated spectrum of the selected samples were evaluated. When the cross-resistance of bacterial strains to different antibiotics was revealed, the correlation coefficient was calculated by Spearman's method [16]. H0 hypothesis for the considered criterion - the value of the correlation coefficient does not differ significantly from zero (there is no reliable correlation). H0 deviated with the estimated probability of its acceptance $P\_$-value $< \alpha = 0.05$. To determine the relationship between MNF and the averaged resistance of the bacterial community to antibiotics, an evaluation using the Spearman correlation coefficient was also used.

3 Results and Its Discussion

The complexity of interrelations in the ecosystem of Lake Baikal, the variability and unpredictability of possible human exposure results in the need for improved microbiological monitoring. Particular attention should be paid to the methods of processing the results obtained. In conditions of anthropogenic pollution of aquatic ecosystems, microbiological monitoring and interpretation of its results acquire features that allow us to identify not only sanitary and epidemiological forecasts, but also the direction of changing microbial communities from general ecological positions. Long-term studies have shown that, in conditions of active anthropogenic pollution of the littoral zone of the Baikal ecosystem, bacterial strains resistant to many antibiotics are observed. When the dependence of the averaged data of antibiotic resistant strains of bacteria on the specific month of sampling was determined and separately by the factor of belonging to a certain year of sampling using the Shapiro-Wilk test, the sampled samples were not distributed according to the normal law ($P\_value < \alpha$). In connection with this, the nonparametric Kraskel-Wallis criterion was used for the variance analysis, and for the correlation analysis the nonparametric correlation coefficient of Spearman. The variance analysis of the average resistance of the bacterial community to antibiotics, grouped by the factor of belonging to a certain sampling month, revealed that in different months of the year the average stability significantly differs from each other ($P\_value = 0.003 < \alpha$). For the estimated time periods from July to November, the minimum value of the average bacterial resistance to antibiotics is observed in July.

Since the height of the tourist season in Lake Baikal during the summer period, this may be due to the introduction of antibiotic-resistant strains into the littoral zone of the lake as a result of anthropogenic activities: massive construction of hotels in the coastal part of the lake, navigation, bathing tourists in the lake, and leaving garbage on the banks. Undoubtedly, the presence of built dachas, cafes, baths, saunas, swimming pools located on the shore and intensively exploited in the summer-autumn period also influences. Uncontrolled drains can directly enter the water with streams flowing into the Baikal or leak from cesspools standing on the shore of cottages. In addition, by August-September, the deep waters of the lake are warming up, the number of storms is increasing, and frequent precipitation is also possible. As a result, over this period of time, flushing from the coastal zone of various types of human activity is observed. Because the water in the lake is cold, bacteria stay in the ecosystem for a long time. The previously accepted view that microorganisms die in the external environment due to low temperatures or lack of nutrient elements is currently undergoing significant changes. Analysis of the actual material obtained in recent years indicates the need to study the role of the external environment as a reservoir of a number of microorganisms. In addition, the peculiarity of the considered ecosystem should be taken into account. Since the ecosystem of Lake Baikal is limited in nitrogen [17], the supply of nitrogenous substances to the lake disrupts biogeochemical processes in the balance of this element.

Analyzing the actual material by season showed that strains of microorganisms resistant to antibiotics can be divided into 2 groups (Fig. 1). The first group: antibiotics with relatively small values of the coefficients of variation in the resistance of bacterial communities to them (coefficient of variation <1). These drugs include: ampicillin, chloramphenicol, neogrammone, trimethoprim. The resistance to antibiotics of this group changes to a lesser extent in the transition from season to season. The second group: antibiotics with relatively high values of the coefficients of variation of the resistance of bacterial communities to them (1 < coefficient of variation <1.75). These are tetracycline, streptomycin, kanamycin, gentamicin, rifampicin, cefazolin, cefatoxime, pefloxacin. The resistance of the bacterial community to the antibiotics of this group is changing to a greater extent, during the five-month period from June to November.

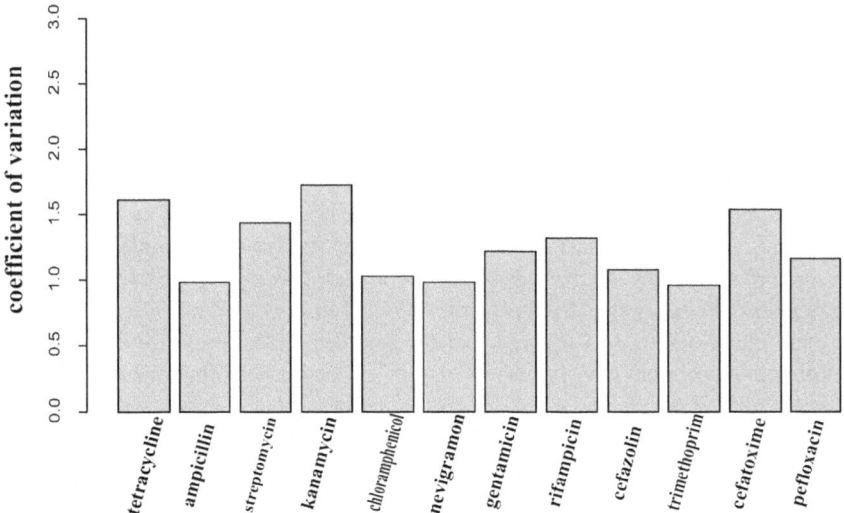

Fig. 1. The value of the coefficients of variation of indicators of bacterial resistance to antibiotics (μg/ml).

It was found that the resistance of bacterial strains to antibiotics varies significantly in different months of the year (P_value = 0.003 < α). Dispersion analysis of the average resistance of bacteria to antibiotics, practically did not reveal the interannual difference in close years (P_value = 0.34 < α). When calculating the pairwise correlation coefficients, it was possible to divide antibiotics into three groups. The first-the stability of pairs of antibiotics is formed independently of each other (values of the correlation coefficient r ≈ 0). The second group consists of pairs with reliable positive values of correlation coefficients ($r > 0$), i.e. the increase in resistance to a single antibiotic was accompanied by an increase in resistance to another antibiotic, the formation of cross-resistance. The third group- these are pairs of antibiotics with reliable negative values of correlation coefficients ($r < 0$). In the framework of the study, an increase in resistance to one drug was accompanied by a decrease in the resistance of bacterial strains to another drug. These are: trimethoprim, an exception with cefazolin, and also cefazolin with streptomycinomas with kanamycin. The value of the correlation coefficient between the total microbial number (MNC) and the average resistance of bacterial communities to antibiotics was 0.075. In this case, the probability of accepting the null hypothesis about the absence of correlation P_-value = 0.41 > α, which indicates the lack of correlation between these indicators. those. UHF and antibiotic resistance are not interdependent. In the subglacial period (January–April), the number of bacteria is lower compared to the period of open water (summer, autumn), but the proportion of bacteria with multiple drug resistance increases. It is possible that this is due to a change in the ratio of autochthonous microflora, subject to seasonal fluctuations in abundance and allochthonous flora, the amount of which can practically not change.

4   Conclusion

The application of the proposed methods in the analysis of a large volume of the results obtained showed that the analysis of samples of bacterial resistance to the antibiotics studied averaged by the factos. Long-term studies have shown that in conditions of active anthropogenic pollution of the littoral zone of the Baikal ecosystem, bacterial strains resistant to many antibiotics are observed. Analysis of samples of bacterial resistance to antibiotics studied averaged by the factor of belonging to a certain sampling month and separately by the factor of belonging to a certain year of sampling using the Shapiro-Wilk test showed that the samples under study are not distributed according to the normal law (P_value < α). In connection with this, the nonparametric Kraskel-Wallis criterion was used for the variance analysis, and for the correlation analysis the nonparametric correlation coefficient of Spearman. The variance analysis of the average resistance of the bacterial community to antibiotics, grouped by the factor of belonging to a certain sampling month, revealed that in different months of the year the average stability significantly differs from each other (P_value = 0.003 < α).

Analysis of the actual material by season showed that strains of microorganisms resistant to antibiotics can be divided into 2 groups. The first group: antibiotics with relatively small values of the coefficients of variation in the resistance of bacterial communities to them (coefficient of variation <1). These drugs include: ampicillin, chloramphenicol, neogrammone, trimethoprim. The resistance to antibiotics of this group changes to a lesser extent in the transition from season to season. The second group: antibiotics with relatively high values of the coefficients of variation of the resistance of bacterial communities to them (1 < coefficient of variation <1.75). These are tetracycline, streptomycin, kanamycin, gentamicin, rifampicin, cefazolin, cefatoxime, pefloxacin. The resistance of the bacterial community to the antibiotics of this group is changing to a greater extent, during the five-month period from June to November. It was found that the resistance of bacterial strains to antibiotics varies significantly in different months of the year (P_value = 0.003 < α).

Dispersion analysis of the average bacterial resistance to antibiotics for the period under review (2005–2006), practically did not reveal the interannual difference in the close years (P_value = 0.34 < α). When calculating the pairwise correlation coefficients, it was possible to divide antibiotics into three groups. The first - the stability of pairs of antibiotics is formed independently of each other (values of the correlation coefficient r ≈ 0). The second group consists of pairs with reliable positive values of correlation coefficients (r > 0), i.e. the increase in resistance to a single antibiotic was accompanied by an increase in resistance to another antibiotic, the formation of cross-resistance. The third group are pairs of antibiotics with reliable negative values of correlation coefficients (r < 0). In bacterial communities for such pairs of antibiotics, an increase in resistance to a single antibiotic was accompanied by a decrease in resistance to another.

References

1. Verhozina, V.A., Kusner, Y.U.S., Safarova, V.A., Sudakova, N.D.: Melkomasshtabnaya turbulentnost' i pehtching bakterioplanktona na Bajkale. Doklady AN. T. **301**(6), 1508–1512 (1988)
2. Verhozina, E.V., Verhozina, V.A., Savilov, E.D., Verhoturov, V.V.: Antibiotikousto-jchivost' mikrobnogo soobshchestva ehkosistemy ozera Bajkal v rajone p. Listvyanka, g. Slyudyanki i g. Bajkal'ska. Byulleten' VSNC SO RAMN. **3**, 62–65 (2014)
3. Anganova, E.V., Samojlova, I.Y.: Antibiotikoustojchivost' bakterij mikrobiocenoza reki Leny v rajone g. YAkutska, Hangalasskogo i Namskogo rajonov (respublika Saha (YAkutiya). Sibirskij medicinskij zhurnal. **7**, 211–212 (2009)
4. Savilov, E.D., Mamontova, L.M., Anganova, E.V., Astaf'ev, V.A.: Uslovno-patogennye mikroorganizmy v vodnyh ehkosistemah Vostochnoj Sibiri i ih rol' v ocenke kachestva vod. Byulleten' VSNC SO RAMN. **1**(129), 47–51 (2008)
5. Anganova, E.V., Anganova, E.V., Rychkova, E.N., Savilov, E.D.: Osobennosti antibiotik-oustojchivosti bakterii mikrobiocenoza reki Vilyuj. Byulleten' VSNC SO RAMN. **2**(54), 72–74 (2007)
6. Anganova, E.V., Anganova, E.V., Savilov, E.D., Savchenkov, M.F., Chemezova, N.N.: G eterogennost' mikrobnyh soobshchestv poverhnostnyh vodoemov po pokazatelyam antibi-otikorezistentnosti bakterij. Gigiena i sanitariya **4**, 19–22 (2014)
7. Savilov, E.D.: EHkologo-ehpidemiologicheskaya ocenka kachestva vod reki Leny. N.: Nauka, 136 (2006)
8. "Pit'evaya voda. Gigienicheskie trebovaniya k kachestvu vody centralizovannyh sistem pit'evogo vodosnabzheniya. Kontrol' kachestva". metodiki MUK 4.2.671–97, razrabotannye SanPiN 2.1.4.559–96
9. Opredelenie chuvstvitel'nosti mikroorganizmov k antibakterial'nym preparatam: Metodicheskie ukazaniya. M.: Federal'nyj centr Gossanehpidnadzora Minzdrava Rossii, 91 (2004)
10. Lakin, G.F. Biometriya, M.: Vysshaya shkola, 349 (1990)
11. Savilov, E.D.: EHpidimiologicheskij analiz: Metody statisticheskoj obrabotki materiala. N.: Nauka-Centr 156 (2011)

12. Amsharin, N.P., Vorob'ev, A.A.: Statisticheskie metody v mikrobiologicheskih issledovaniyah. L.: Medgiz, 180 (1962)
13. Kraemer, H.: Correlation coefficients in medical research: from product moment correlation to the odds ratio. Stat. Methods Med. Res. **15**, 525–544 (2006)
14. Grzhibovskij, A.M.: Korrelyacionnyj analiz. EHkologiya cheloveka. No 9 50–60 (2008)
15. Royston, P.: Algorithm AS 181: The W test for Normality. Appl. Stat. **31**, 176–180 (1982)
16. Hollander, M.: Nonparametric Statistical Methods. In: Hollander, M and Wolfe, D.A. (eds.) Wiley, New York (1973)
17. Verhozina, V.A., Verhozina, E.V., CHudnenko, K.V.: Rol' biogeohimicheskih processov v balanse azota ehkosistemy ozera Bajkal . Voda: himiya i ehkologiya. **12**, 3–7 (2011)
18. Verhozina, V.A., Verhozina, E.V., Verhoturov, V.V., Safarov, A.S.: Monitoringovye issledovaniya mikrobnogo soobshchestva litoral'noj zony v rajone yuzhnogo Bajkala. Voda: himiya i ehkologiya. **3**, 66–70 (2014)

Mapping of Model Estimates of Phytoplankton Biomass from Remote Sensing Data

Svetlana Ya Pak$^{(\boxtimes)}$ (ID) and Alexander I. Abakumov (ID)

Institute of Automation and Control Processes, Far Eastern Branch of the Russian
Academy of Sciences, 5 Radio Str., Vladivostok 690041, Russia
{packsa, abakumov}@iacp.dvo.ru

Abstract. Phytoplankton is the lowest level of the trophic chain determining
the aquatic ecosystem productivity. Information about the surface phytoplankton
distribution over a large area can be obtained by the modern remote methods.
Satellite signal penetrates only into the upper layer, so these methods are limited.
Plant biomass volume located under the surface water layer differs significantly
from the remote data. Vertical model of phytoplankton functioning based on the
concept of fitness function is used to reconstruct the integral biomass in the
whole water column under a unit area. Phytoplankton community is considered
under its aspiration to occupy the niche most favorable for life. The community
growth rate coincides with the specific growth rate of phytoplankton. The model
solution reduces to solving the Cauchy problem for a system of ordinary differential equations with the remote sensing data as the initial conditions. The
remote sensing data of the Sea of Japan and Issyk-Kul Lake are used for the
model testing. The model solution visualization gives an idea of the spatial
distribution of biomass within the entire zone where the photosynthesis takes
place.

Keywords: Mathematical model · Phytoplankton · Satellite · Remote data ·
Primary production · Assimilation function

1 Introduction

The rapid development of rapid satellite monitoring techniques enables the visualisa-
tion of chlorophyll distribution on the surface of the world's oceans. However, the
main biomass of phytoplankton is often quite far below the surface layer [1], as
confirmed by numerous measurements in situ [2, 3]. The current paper suggests a
method for mapping the integrated biomass of phytoplankton within a given geographic object using mathematical modelling methods.

2 Vertical Biomass Model

Phytoplankton is assumed to function in a fixed water column under a surface unit. Its
biomass estimates are based on a stationary model of phytoplankton functioning as
follows:

© Springer Nature Switzerland AG 2019
I. Bychkov and V. Voronin (Eds.): *Information Technologies
in the Research of Biodiversity,* SPEES, pp. 73–79, 2019.
https://doi.org/10.1007/978-3-030-11720-7_11

$$\frac{dy}{dx} = \left[\frac{d\mu}{dx} - e(y)\right]y, \quad \frac{dz}{dx} = vp(y, y_0)z, \quad \frac{dI}{dx} = -k(y, z)I. \tag{1}$$

where $y(x)$ is the phytoplankton biomass density (g/m^3), z (t, x) is the nutrient concentration (g/m^3) and $I(x)$ is the light irradiance of the ocean surface and its depth distribution.

The main model assumptions are as follows:

1. Phytoplankton functioning takes place at a fixed time in a static water column. Hydrological effects are levelled.
2. Fitness function [4, 5] coincides with the specific growth rate of the phytoplankton biomass $\mu(z, I, \theta) = \mu_0 \cdot \mu_z(z) \cdot \mu_I(I) \cdot \mu_\theta(\theta)$, which is interpreted as the community growth rate.
3. A reverse effect is observed between phytoplankton and nutrient concentration (food stimulus).

Let us consider in greater detail the quantities occurring in the equations of system (1).

First Equation

The coefficient μ_0 is the maximum possible phytoplankton growth rate, $\mu_z(z) = z/(z_0 + z)$, where z_0 is the half saturation coefficient of biogenic nutrition; $\mu_I(I) = I/(I_0 + I)$, where I_0 is the half saturation coefficient of the light irradiance; and $\mu_\theta(\theta) = \exp\left(-\frac{(\theta - \theta_{opt})^2}{2\tau^2}\right)$, where θopt is the aquatic environment temperature optimal for phytoplankton growth, τ is the tolerance interval, and e (y) is the specific rate of the phytoplankton biomass elimination that occurs because of zooplankton eating and other causes of mortality.

Second Equation

Assumption 3 is $p(y, y_0) = \frac{y_0}{y + y_0}$, where y_0 is the half saturation coefficient for phytoplankton, and v is the maximum possible rate of the nutrient replenishment [6].

Third Equation

This ratio simulates the illumination attenuation in depth caused by the refraction of water layers, shading by phytoplankton and suspended organic matter. The corresponding distribution is described by the function $k(t, x, y, z) = k_0 + k_1 y + k_2 z$, where k_0 is the total water turbidity coefficient, k_1 is the phytoplankton shading coefficient and k_2 is the mineral substance shading coefficient [7].

System (1) can be verified in several ways depending on the quantity and quality of information the researcher has. In this study, the regulatory factors are modelled as external functions using a biogeographical description of a preset object. Another approach is possible. Optimisation methods, for example, least squares method, are recommended if contact measurements are available. Regardless of the method, the objective is to use all available information to maximise the possible refinement of model parameters.

3   Use of Remote Sensing Data

Model (1) is a system of ordinary differential equations. We use biological information on the state of the upper (up to 10 m) water layer as the initial conditions. A uniform grid with a certain resolution is applied to a given geographic region. The response of the satellite signal is recorded at its nodes. After an appropriate interpretation [8], the specific concentrations of chlorophyll, surface temperature and illumination can be used to calculate the integrated biomass at any known point by solving system (1). Relevant data are provided by the Centre for Regional Satellite Monitoring of Environment (IACP FEB RAS, Vladivostok).

Realisation of the vertical model requires knowing the geographical depth corresponding to the given coordinates. The depth map can be constructed based on the topographic material, for example, by digitising a raster image of a map with a known scale. Modern software can convert the obtained isolines into a regular grid, coordinated with satellite data. Digitisation and transformation of topographic information are carried out in the computer graphic laboratory IACP FEB RAS using Vextractor (http://www.vextrasoft.com) and Surfer 14 (Golden Software http://www.goldensoftware.com).

4   Sea of Japan

One of the objects of the constructed model (1) application is the Sea of Japan, specifically, the section from $127°$ to $142°$ east longitude and from $34°$ to $47°$ northern latitude.

Based on the biogeographic differentiation of the Sea of Japan [9], temperature is defined as an external regulating factor using the following function:

$$
\begin{cases}
\theta_0(t), & 0 \leq x \leq x_1^\theta(t,\eta) \\
\varphi\left(x; x_1^\theta(t,\eta), x_2^\theta(t,\eta), \theta_0(t), \theta_b\right), & x_1^\theta(t,\eta) \leq x \leq x_2^\theta(t,\eta) \\
\theta_b, & x_2^\theta(t,\eta) \leq x \leq \bar{x}
\end{cases}
$$

where $\theta_0(t)$ is the surface temperature.

To determine the seasonal thermocline boundaries, we introduce the function $\varphi\left(x; x_1, x_2, y_1, y_2\right) = y_1 + (y_2 - y_1)\frac{x-x_1}{x_2-x_1}$ and the variable $\eta = \lambda - \zeta$, where λ and ζ are the longitude and latitude of the studied point in the Sea of Japan, respectively.

The temperature at depth = 100 m is denoted as θ_b and defined as follows:

$$\theta_b(t, \eta) = \varphi(\eta; \eta_1, \eta_2, \varphi(min\{t, 5/3\,(T - t)\}; 0, 5/8\,T, 0, 2), 8),$$

where $T = 365$ - full year (days).

The values of η_1 and η_2 correspond to the northwestern (the smallest possible value of η) and southeastern points (the largest possible value of η), respectively. The peak of summer temperatures is in August. The temperature change boundaries in depth are described by the following functions:

$$x_1^\theta(t, \eta) = \varphi(\eta; \eta_1, \eta_2, \varphi(min\{t, 5/3\,(T - t)\}; 0, 5/8T, 0, 15), 0),$$

$$x_1^\theta(t, \eta) = \varphi(\eta; \eta_1, \eta_2, \varphi(min\{t, 5/3(T - t)\}; 0, 5/8\,T, 100, 50), 100),$$

$$\theta_{opt}(\eta) = \varphi(\eta; \eta_1, \eta_2, 5, 15), \ \tau = 15.$$

Thus, we determine the function of the temperature change in depth depending on the season and the geographical location of the observed zone of the Sea of Japan. The values of other parameters of the system are taken in accordance with Tanaka and Mano [10] and Abakumov and Izrailsky [11].

5 Primary Production

Model (1) enables to calculate the integrated phytoplankton mass in the entire water column. If the proportion of chlorophyll is known, then defining the integrated chlorophyll mass and subsequently building some estimates of primary production in the entire water volume are possible. The following generally accepted concept is used to solve this problem:

$$P = B \times P^B,$$

where P is the primary production per unit volume, B is the chlorophyll mass and P^B is the assimilation function. We used Platt's dependence [12] to calculate P^B and data from [13] to refine the parameters of the assimilation function.

$$P^B(t, z, \theta, I) = \psi(t)P_m^B(z, \theta)\left[1 - \exp\left(-\alpha^B I/P_m^B(z, \theta)\right)\right],$$

$\Psi(t)$ - the light day duration (h),

$$P_m^B(z, \theta) = P_r^B \cdot g(z) \cdot f(\theta), g(z) = \mu_z(z), f(\theta) = \exp(k\theta), \quad k = 0.12$$

$$P_r^B = 0.84\left(gC \cdot gChl^{-1} \cdot h^{-1}\right)$$

$$\alpha^B = \alpha_r^B \cdot g(z) \cdot f(\theta), \quad \alpha_r^B = 0.08854\left(gC \cdot (gChl)^{-1}h^{-1}\left(Em^{-2}day^{-1}\right)^{-1}\right).$$

Estimates of the total annual production in the entire water area available to satellite observation are presented in Fig. 1.

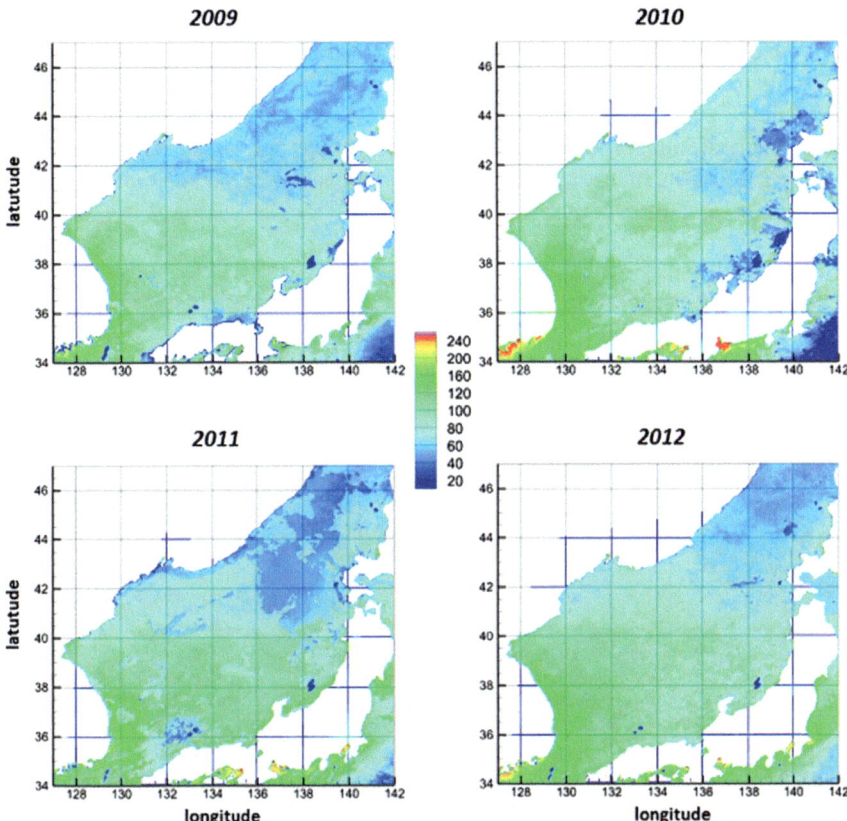

Fig. 1. Estimates of the annual primary production of the Sea of Japan (gC $\cdot$ m^{-2}year^{-1}).

6   Issyk–Kul Lake

The study object is a geographic area from 77° to 78° east longitude and from 42° to 43° northern latitude. The data are obtained during the expedition study on Issyk–Kul Lake [14] in the Tamga–Grigorievka pelagic section.

Phytoplankton biomass, temperature, light irradiance and nutrient concentrations recorded at different horizons are approximated by the spline interpolation method to further minimise the residual functional. The standard deviation of the left and right parts of the first equation of the system (1) is minimised with respect to parameters $\mu_0, z_0, I_0, \theta_{opt}$ and τ:

$$\left[\int_0^{x_b} \left(\frac{dy}{dx} - \left[\frac{d\mu}{dx} - e(y) \right] y \right)^2 dx \right]^{1/2} \tag{2}$$

The integration limits are chosen in accordance with the horizons on which the reference values of all the constructed splines are known. The spline function approximating the vertical light irradiance distribution is limited to a depth of $x_b =$ 75 m, which determines the upper limit of integration.

The obtained values then become the main parameters of model (1). The other parameters are taken from Tanaka and Mano [10] and Abakumov and Izrailsky [11]. With satellite information used as the initial conditions for the system (1), the mean daily estimates of the integrated phytoplankton biomass were obtained in July–August 2002 from the cloudless portion of the Tamga–Grigorievka section (Fig. 2).

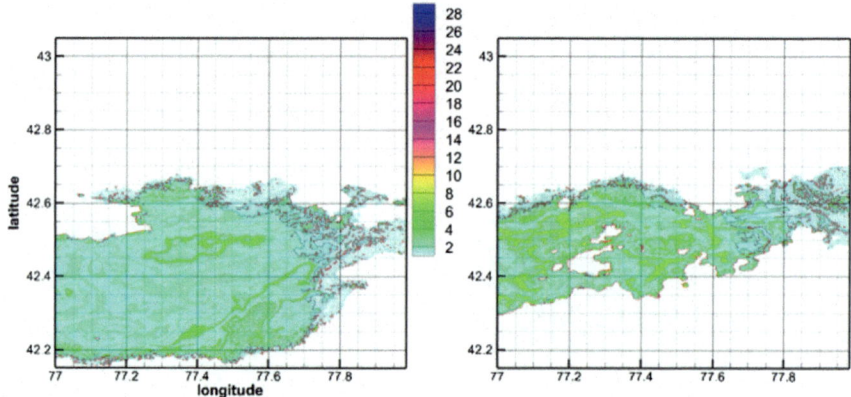

Fig. 2. Integral estimates of phytoplankton biomass (mg/m^3) for July (left) and August (right) 2002.

The maximum estimates of the integrated phytoplankton biomass in the Issyk–Kul Lake are in complete agreement with the measurements presented in Tynybekov and Matorin [15]. The calculation of model primary production is comparable with those of similar studies published in the press [16]. This result attests to the suitability of model (1) for constructing predictive estimates of the total biomass and primary production in any geographic area available for satellite processing.

Better results can be obtained if contact data are collected systematically in an observed region and satellite data processing algorithms are improved.

Acknowledgments. The reported study was funded by RFBR according to the research project № 18-01-00213.

The reported study supported by a grant of Comprehensive program of fundamental scientific research « Far East » (project № 18-5-051).

References

1. Lepskaya, E.: Distribution of phytoplankton in the water area of the northern part of the West Kamchatka shelf in the middle of summer 2008. *Issledovanija vodnyh biologicheskih resursov kamchatki i severo-zapadnoj chasti tihogo okeana* [Research of aquatic biological resources of Kamchatka and the northwestern part of the Pacific Ocean] **42**, 71–77 (2016) https://doi.org/10.15853/2072-8212.2016.42.71-77. (in Russian)

2. Finenko, Z., Churilova, T., Lee, R.: Dynamics of the vertical distributions of chlorophyll and phytoplankton biomass in the black sea. Oceanology **45**(1), 112–126 (2005)
3. O'Reilly, John, E., Zetlin, C.: Seasonal, horizontal, and vertical distribution of phytoplankton chlorophyll a in the Northeast U.S. Continental Shelf Ecosystem. Seattle, WA, NOAA/National Marine Fisheries Service, (NOAA Technical Report NMFS, 139) (1998)
4. Polujektov, R.A., Pyh, J.A., Shvytov, I.A.: *Dinamicheskie modeli jekologicheskih sistem* [Dynamic models of ecological systems]. Gidrometeoizdat, Leningrad (1980). (in Russian)
5. Frisman, E., Zhdanova, O.: Evolutionary transition to complex population dynamic patterns in a two-age population. Russian J. Genet. **45**(9), 1124–1133 (2009). https://doi.org/10.1134/s1022795409090142
6. Pak, S., Abakumov, A.: A model method for restoring the state of phytoplankton in the vertical water column from satellite data on the surface layer. *Informatika i sistemy upravlenija* [Informatics and Control Systems] **3**(41), 23–32 (2014). (in Russian)
7. Abakumov, A., Izrailsky, Yu., Park, S.: Functioning of the phytoplankton in seas and estimates of primary production for aquatic ecosystems. Dev. Environ. Model. **27**, 339–349 (2015). https://doi.org/10.1016/b978-0-444-63536-5.00015-6
8. Aleksanin, A., Kim, V., Orlova, T. et al.: Phytoplankton of the Peter the Great Bay and its remote sensing problem. Oceanology **52**(2), 219–230 (2012). https://doi.org/10.1134/s0001437012020014
9. Dobrovolsky, A., Zalogin, B.: The Seas of the USSR. MGU Publ., Moscow (1982)
10. Tanaka, Y., Mano H.: Functional traits of herbivores and food chain efficiency in a simple aquatic community model. Ecol. Model. **237–238**, 88–100 (2012). https://doi.org/10.1016/j.ecolmodel.2012.04.021
11. Abakumov, A., Izrailsky, Y.: Environment influence on the phytoplankton distribution in a basin. Math. Biol. Bioinf. **7**(1), 274–283 (2012)
12. Platt, T., Caverhill, C., Sathyendranath, S.: Basin-scale estimates of oceanic primary production by remote sensing: the North Atlantic. J. Geophys. Res.: Oceans (1978–2012) **96**(8), 15147–15159 (1991)
13. Grangere, K., Lefebvre, S., Menesguen, A., Jouenne, F.: On the interest of using field primary production data to calibrate phytoplankton rate processes in ecosystem models. Estuar. Coast. Shelf Sci. **81**, 169–178 (2009)
14. Matorin, D., Antal, T., Sharshenova, A., et al.: Study of natural phytoplankton of the Issyk-Kul lake obtained using an immersion fluorometer. *Vestnik MGU* [Bulletin of the Moscow State University]. Ser. Biol. **1**, 43–45 (2002)
15. Tynybekov, A., Matorin, D.: The State of Phytoplankton of the Issyk-Kul Lake. KRSU Publishing, Bishkek (2009)
16. Fu, G., Baith, K., McClain, C.: SeaDAS: The SeaWiFS data analysis system. In: Proceedings of the 4th Pacific Ocean Remote Sensing Conference, pp. 73–79. Qingdao, China (1998)

About the Project of the Web GIS "Electronic Atlas of Bryophytes of the Republic of Bashkortostan"

T. U. Biktashev⬛, N. I. Fedorov⬛, and E. Z. Baisheva$^{(\boxtimes)}$⬛

Ufa Institute of Biology—Subdivision of the Ufa Federal Research Centre of the
Russian Academy of Sciences, October av., 69, Ufa 450054, Russia
`biktashev.timur@rambler.ru`, `fedorov@anrb.ru`,
`elvbai@mail.ru`

Abstract. The open information and analytical web geographic information system "Electronic Atlas of Bryophytes of the Republic of Bashkortostan" is being developed now to improve the sharing bryological data between the botanists from Ufa Institute of Biology and other research groups. Open source software was used for development the application. LeafletJS is used for visualized spatial data, it provides fast creating interactive maps, refresh images, images are composed from tiles and changing map scale. ASP.NET Core is used as platform for web-application. Sources may be found in GitHub. The primary cartographic data are stored in DBMS PostgreSQL with extension PostGIS. PostgreSQL have multiuser access, it is open source project and has capability to work in many operating systems. Geoserver is used for visualized vector and raster data. Layer styling uses Styled Layer Description (SLD) which certificated by Open Geospatial Consortium (OGC). All cartographic data was prepared with QGIS. Biodiversity protection strategies, designed by sharing information and integrating data, play an important role in defining interconnections in research as well as in increasing global awareness. We hope that developed GIS application will improve the monitoring and protection of bryophyte diversity in the Republic of Bashkortostan.

Keywords: Bryology · Biodiversity · Web services · Information system · Southern ural

1 Introduction

The investigations of spatial-temporal dynamics of biodiversity and distribution patterns of species are priority tasks of fundamental ecology and biogeography. An important tool for monitoring biodiversity is a geographic information system (GIS), which accommodates large varieties of spatial and attribute data. The creation of digital maps and databases on flora and vegetation, as well as the use of special GIS-programs with automation of the process of botanical data analysis are examples of the modern methods using in scientific research and protecting the environment and plant diversity in general [1, 2].

© Springer Nature Switzerland AG 2019
I. Bychkov and V. Voronin (Eds.): *Information Technologies
in the Research of Biodiversity,* SPEES, pp. 80–85, 2019.
https://doi.org/10.1007/978-3-030-11720-7_12

GIS can be used as an effective tool for monitoring biodiversity changes, application of mathematical methods in the study of relationship between the distribution and habitat conditions of species, as well as evaluation of relationship between environmental factors and species range, for prediction of potential species range, etc. Usually, database provides standardized information about diversity of species, and includes names, species abstracts, regional check-lists, precise distributional data, information on regional ecology and population size of species, bibliographic references and images. It can be used as the basis for biodiversity monitoring, for species conservation planning, the creation and zoning of nature protected areas. These data are necessary for mathematical modeling, predictions of the human impact on the habitats of species, as well as the impact of climate change on biodiversity and ecosystems. The creation of information systems allows optimizing costs for scientific and environmental purposes and reduce government spending by minimizing duplication of research and making information exchange possible across organizations and research teams [3, 4].

The history of bryological investigations on the territory of Bashkortostan is not very long. The first data about bryophytes collected on this territory were published at the end of 19th century [5]. Nevertheless, the bryological research of last decades allowed to obtain a large amount of bryological data from numerous protected areas (three nature reserves, one national park, some nature monuments) and to reveal bryophyte diversity of the main habitats and vegetation types of study area. Similar to many other Russian Regions, the Bashkortostan has a long record of botanical data (herbarium specimens, geobotanical relevés, scientific reports, Bachelor's and Ph.D. thesis, etc.) that are not easily accessible to a broad range of scientists. The ongoing project is aimed to provide all bryological data from Bashkortostan to a wide range of specialists in the online portal.

At present, the authors are developing information and analytical web geoinformation system on bryophytes of the Republic of Bashkortostan. Modern web applications are built on service oriented architecture, web applications are provided as complex of interrelated software for managing data – create, post, edit, compute, display and distribute. The basis of geoinformation web application is usually the libraries of spatial data visualization like OpenLayers, LeafletJS, Yandex maps and etc. For the development web application "Electronic Atlas of Bryophytes of the Republic of Bashkortostan" is used LeafletJS library. This library will provide fast creating interactive maps, fast refresh images, images are composed from tiles and changing map scale. The Library will provide access for map view functions: map visualization, thematic map data, for example: distribution of species within specially protected natural areas, possibility of output of primary information about data (name of collector or observer, date of collecting, type of surrounding vegetation, etc.).

2 GIS Functional Structure

GIS functional structure is shown in Fig. 1. The subsystems are as follows.

Input/output subsystem – digital maps, table data information of the location of species with geographic coordinates, the source of information (herbarium specimens, geobotanical relevés, literature data, etc.), habitat type, habitat type in the EUNIS system, etc.

Reference information subsystem – includes a list of species related with families and genera, authorship of species, list of EUNIS habitats with the description, a list of alliances of floristic classification.

Spatial and attributive data storage subsystem – it uses open source multi-user DBMS PostgreSQL with PostGIS module for storage spatial data.

Data processing and analysis subsystem - provides statistical processing and output data of variety of species in grid cells, quantitative representation of taxonomical categories, the ability of selection information received for a certain period of time from any collectors.

The preparation and presentation of output data subsystem – exports diagrams and table data with primary attribute data.

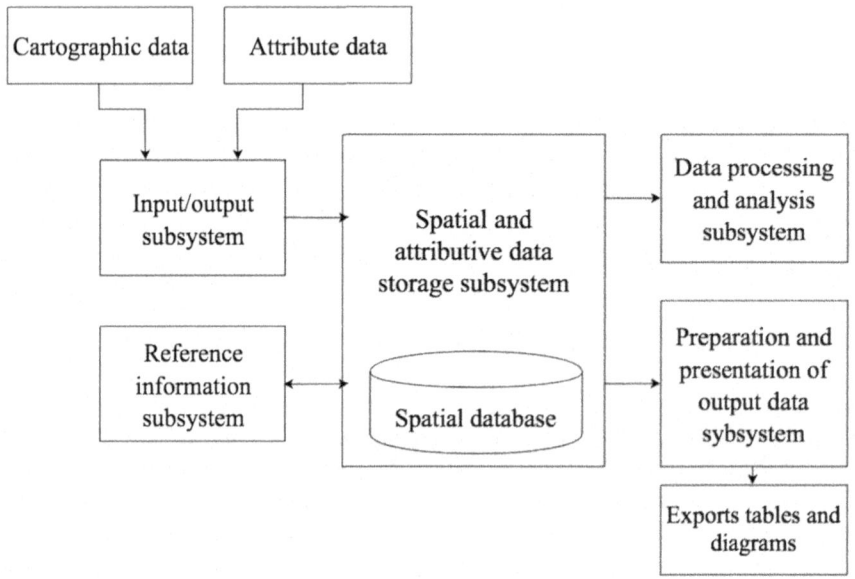

Fig. 1. GIS structure.

3   GIS Architecture

The architecture of distributed GIS applications consists of: a client side web-application, a service of processing spatial requests and a map server. Client side GIS application provides managing spatial and attribute data interface, makes diagrams, exports primary data information and maps which includes information about distri-bution of each bryophyte species known from the Republic of Bashkortostan. The processing service of attributive data uses requests to execute operations with attribute data.

The cartographic service is designed to display spatial data. Service oriented architecture is provided to use web-application on the client side computers which supports the standard web-browsers. The main computing is performed on the server side. All computing is performed on server and as a result the amount of data transmitted is gone down. Eventually, an end user receives only the results of computing request data, not all data necessary for its execution. Thus, the architecture of GIS application is illustrated on Fig. 2.

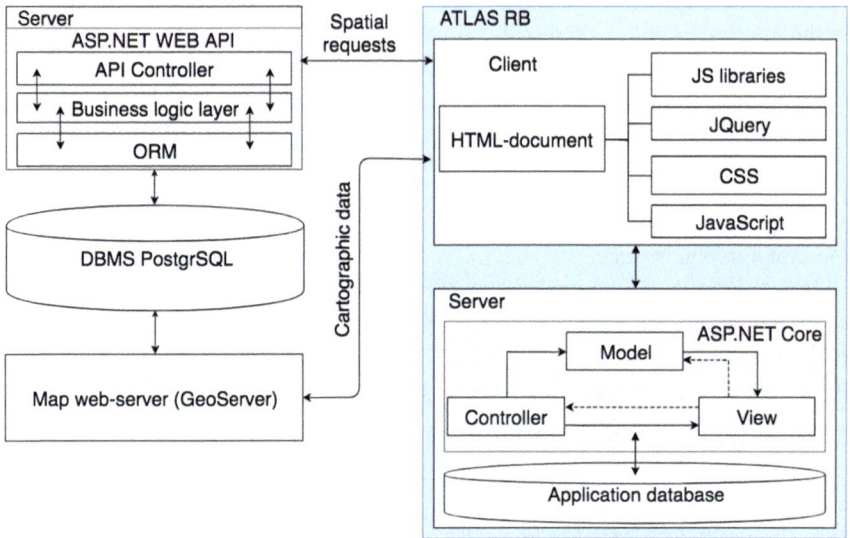

Fig. 2. GIS of "Electronic Atlas of Bryophytes of the Republic of Bashkortostan" architecture.

4 Development

The open source software was used for development the application: ASP.NET Core, Geoserver, PostgreSQL with extension PostGIS. LeafletJS is used for visualized spatial data, it was chosen for compact size and ease of development [6]. ASP.NET Core was used as platform for web-application. ASP.NET Core [7, 8] is an open source framework from Microsoft, it is designed for building different web-applications. Sources may be found in GitHub. The primary cartographic data is stored in DBMS PostgreSQL [9] with extension PostGIS [10]. PostgreSQL has multiuser access, it is open source project and designed to work in many operating systems. Geoserver [11] is an open source software, it visualizes vector and raster data. Layer styling uses Styled Layer Description (SLD) which is certificated by Open Geospatial Consortium (OGC). All cartographic data was prepared with QGIS.

At present, web GIS "Electronic Atlas of bryophytes of the Republic of Bashkortostan" in which all current data about bryophyte flora of the Republic of Bashkortostan will be submitted is being developed and tested. The functionality includes visualization of data at the location of each species in the form of rectangular cells

10 × 11 km in size and 1.0 × 1.1 km in size, which, depending on the scale of the display, will be displayed separately or together on loadable on-demand cartographic substrates (OpenStreetMap, map of the territorial division of the Republic of Bashkortostan, map of the nature protected areas of the republic, etc.). Species search in a database will be carried out by input name or selection from the list of all bryophytes revealed in the territory of republic. After selection "Species", it will load page with information about the name of the species (Latin name, synonyms and Russian name), its image, characteristics of habitat, species distribution map on the territory of the Republic of Bashkortostan at a scale of 1:5000000 (1 cm = 50 km), with the species localities marked by square cells with dimension 10 × 11 km. The page will contain two graphic buttons - "Distribution" and "Ecology". After clicking the "Distribution" button, three layers will be loaded - two polygon layers, where the species location will be presented by square cells 10 × 11 km and 1.0 × 1.1 km, and a raster layer - the basemap (by default – OpenStreetMap). After clicking the square cell – species location will be displayed information like habitat (description of surrounding plant community, habitat type in EUNIS), the date of the species finding and reference to the bibliographic source.

After clicking the "Ecology" button, you will be redirected to the help page, where you can find information about species presence in various types of EUNIS habitats, vegetation communities (the alliances of floristic classification according Braun-Blanquet approach) and for rare species – reasons of rarity and the protection status.

5   Conclusion

Ecology conservation strategies designed by sharing information and integrating data, play important role in defining interconnections in research. The creation of a GIS for the Russian region, which combines database and small-scale digital maps of distribution species with large-scale and location linking will promote research of distribution patterns of bryophytes, ecological and phytosociological role of species in different parts of range, and integrate data on bryophytes of Bashkortostan into Russian and international databases providing this data to a wide range of specialists. The created GIS application will improve the monitoring and protection of bryophyte diversity in the Republic of Bashkortostan. In the future, this application may easily be adapted for both research of other taxonomic groups of plants, and for other regions.

Acknowledgements. Authors are grateful to Russian Foundation of Basic Research (grant no. 18-04-00641/18) for the financial support of this research.

References

1. Salem, B.B.: Application of GIS to biodiversity monitoring. J. Arid Environ. **54**, 91–114 (2003). https://doi.org/10.1006/jare.2001.0887
2. Prasad, N., Semwal, M., Roy, P.S.: Remote sensing and GIS for biodiversity conservation. In: Upreti, D.K. et al. (eds.) Recent Advances in Lichenology, pp. 151–179, Springer India (2015). https://doi.org/10.1007/978-81-322-2181-4_7
3. Chapman, A.D., Muñoz, M.E.S., Koch, I. Environmental information: placing biodiversity phenomena in an ecological and environmental context. Biodivers. Inform. **2**, 24–41 (2005)
4. USDA, NRCS Homepage, https://plants.usda.gov/about_plants.html. Last accessed 08 Aug 2018
5. Шелль, Ю.К. Материалы для ботанической географии Уфимской и Оренбургской губерний (Споровые растения). Труды Общества естествоиспытателей при Императорском Казанском университете [Shell, Yu.K. Materials for the botanical geography of Ufa and Orenburg provinces (Cryptogamic plants). Trudy obshchestva estestvoispytateley pri Imperatorskom Kasanskom universitete] **12**(1), 1–93 (1883)
6. LeafletJS HomePage, https://leafletjs.com/. Last accessed 08 Aug 2018
7. Introduction to ASP.NET Core, https://docs.microsoft.com/en-us/aspnet/core/?view=aspnetcore-2.1. Last accessed 08 Aug 2018
8. Freeman, A.: Pro ASP.NET Core MVC 2. 7th edn. Apress, London UK (2017)
9. PostgreSQL Documentation, https://www.postgresql.org/docs/. Last accessed 08 Aug 2018
10. PostGIS HomePage, https://postgis.net/. Last accessed 08 Aug 2018
11. Geoserver About, http://geoserver.org/about/. Last accessed 08 Aug 2018

Bioclimatic Data Optimization for Spatial Distribution Models

Mikhail Orlov[1]([⊠]) and Alexander Sheludkov[2]

[1] Institute of Cell Biophysics of Russian Academy of Sciences, Pushchino,
Russia
orlovmikhailanat@gmail.com
[2] Institute of Geography, Russian Academy of Sciences, Moscow, Russia

Abstract. Spatial distribution models (SDMs) are successfully used across various aims of biology, ecology, environment protection, etc. as means to predict distribution areas of living species. This includes changes in the distribution upon environmental changes, invasions, and other dramatical alterations affecting both biota and humans. For the purpose of SDMs training Maximization of Entropy (Maxent) machine learning algorithm is most applicable one. As for predictors set, climatic variables are among widely used. Numerous works addressing the problem in general have shaped the commonly used workflow. Here we consider the possibility to expand the workframe by applying unsupervised machine learning techniques (clusterization, PCA, and correlation analysis) fon input SDMs data for their optimization as well as exploration of the bioclimatic dataset. The need is connected to the fact that highly correlated predictors and excessively large data are likely to decrease machine learning performance. Having obtained the list of less contributing variables, we derived the new reduced dataset from the initial one by removing predictors from the list. Both datasets served as predictor sources for training of classifiers based on various machine learning methods. This allowed to produce better performance for some methods including Maxent while having dataset size decreased. Additionally, good agreement was evidenced for distribution areas predicted by Maxent and by the rest algorithms used, which implies that their simultaneous usage might help better robustness.

Keywords: SDM · Crimea · Ecoligical modeling · Data optimization

1 Introduction

As the body of data in geographical and biological sciences grows, the need in more accurate and efficient ways of automated processing and analysis increases accordingly. Among these methods species distribution models (SDMs), which are powerful and versatile tools to assess the interconnection between species localities and the environmental as well as other characteristics of corresponding sites [1]. Since 2000s SDMs are gaining growing attention of experts representing wide range of fields including biogeography, botany, ecology, etc. [2]. Significant advances were achieved in the field yielding a plethora of methods and implementations. The distinctions among them primarily in type of data on species distribution used. In case of common in ecology

© Springer Nature Switzerland AG 2019
I. Bychkov and V. Voronin (Eds.): *Information Technologies
in the Research of Biodiversity,* SPEES, pp. 86–95, 2019.
https://doi.org/10.1007/978-3-030-11720-7_13

systematically collected data (including formal biological surveys where a set of sites are checked and either the presence/absence or abundance of species are established) regression modeling methods have attracted the most attention. These include generalized linear models (GLM), additive linear models (GAM), random forests, etc.

In most cases, however, systematic surveys information is not available or truncated. Indeed, species records are mostly comprise presence-only records originating, for example, from herbarium and museum collection tags. Hence, this sort of data holds the majority of information in the problem field and demands thorough analysis. Currently various SDM methods for modelling presence-only data are available, from which MaxEnt clearly stands out [3] given that its predictive capacity is comparable to models performing the best [4]. Since elaborated in 2004, Maxent has been applied extensively for modelling species distributions. Spectrum of works that uses the methods covers various aims of biology, evolution studies, as well as conservation and biosecurity applications.

Various data describing features of sites or territories could be used as input data for Maxent algorithm with climate characteristics data being among most popular. These variables are often of similar nature and therefore prone to correlate, especially in case of neighbouring sites. General practice of machine learning have pointed that highly correlated predictors are likely to impede classifier performance. Indeed, in case of a large number of predictors or predictors that are highly correlated only a subset of the variables is necessary for the construction of 'sufficient predictors' [5]. As for a set of climatic or bioclimatic variables, some of them are bound to be correlated by their definition (e.g., precipitation of wettest month and precipitation of wettest quarter, etc.) For example, certain bioclimatic parameters stored in the Worldclim database [6] are merely the products of division of one variable by another or different type of transformation including basic arithmetic operations. This makes an initial dataset inflated and containing statistical noise, which poses a need in dimensionality reduction as well as optimization. In such cases, the variables removent and not replacement, engineering, etc. appear to be reasonable. The task could be performed using various approaches, in particular, by means of unsupervised machine learning techniques. As a side note, at the stage of the workflow one can expect that new variables engineering might also be beneficial; however, this is in the scope of our future work.

As for machine learning algorithm selection, two decade-long experience of Maxent usage makes it the solid tool for spatial distribution modeling. Its advantages are justified in multiple work, for example [7]. In particular, it enables quantifying the importance of the variables with percent contribution and permutation importance values based on the readily implemented permutation test [8]. This however requires having the model already built and thus rather describes one particular modeling result and not the data structure. Unfortunately, applying permutation test with machine learning methods beside Maxent requires workaround due to the absence implementations that could be applied uniformly. Accordingly, obtaining common techniques enabling variables importance assessment for Maxent as compared to different techniques would be beneficial. Purely data-driven approach used prior to supervised machine learning step is one attractive possibility here. It additionally enables data exploratory analysis, validation, and optimisation. To sum up, one can suggests augmenting Maxent-based approach with certain other techniques for the advancement of

SDMs training, in particular, in order to enhance given framework efficiency and robustness. Here we report data optimisation and the preliminary comparison of Maxent to various different machine learning techniques (representing boosting algorithms, decision trees ensemble, naive bayes algorithm, neural networks, etc.) as implemented completely in R free statistical software [9].

2   Case Region

The Crimean peninsula is located between 44°23′N and 46°14′N, 32°28′E and 36°39′E, surrounded by the Black Sea to the west, south and southeast and by the Sea of Azov to the northeast. Most of the territory of Crimea represents plains, while mountains cover the southern and southeastern parts of the peninsula. The Crimean Mountains consist of three parallel ridges, stretched from the Southwest to the Northeast up to 160 km. The highest one is the Southern (the Main) ridge, with maximum heights of more than 1.5 thousand meters, gentle northern slopes and steep southern slopes.

For most of the year the peninsula is experiencing an intense influx of solar radiation. The atmospheric circulation is characterized by the prevalence of westerlies, which cause a constant air inflow from the Atlantic. The precipitation amount varies from 250 mm per year in steppe areas to 1000 mm per year in the mountains. In wintertime one can observe a significant warming effect of the sea on the coastal areas. The number of frosty days per year here is on average half as much as in inner parts of the peninsula. The southern coast is also protected by the mountains from the invasions of cold air masses from the north. As result, in this region the average monthly air temperature remains positive throughout the year. Thus, generally the territory of the Crimean peninsula is divided into three climatic regions: the plains with temperate warm and arid type of climate, the mountains with temperate warm and humid climate, and the southern coast with subtropical Mediterranean climate [10].

3   Materials and Methods

Worldclim database (Version 2.0) with a spatial resolution of 2.5 arc-minutes (approx. 5 km) [6] served as a source for the initial set comprising 19 bioclimatic variables. The set was used for exploratory analysis, data optimization, and later on as predictors for SDMs training. The variables' values were extracted for a rectangular grid with resolution of 7′ 30″ (approx. 0.125°), which produced 837 points in total. Actual plant localities were obtained from the collection of georeference tags from three Russian herbariums (Moscow University Herbarium (MW), Main Botanical Garden of Russian Academy of Sciences (MHA), and Herbarium Russian Academy of Sciences - V. L. Komarov Botanical Institute (LE) that originated from the Crimean peninsula. The most numerous plant (131 locality points) was the representative of the Apiaceae family *Pimpinella tragium* (see the corresponding distribution on the map on Fig. 1). The relatively large number of observations for the species allowed to improve performance of models (models for second and third most presented plants were shown to perform poorly and therefore excluded from the subsequent consideration).

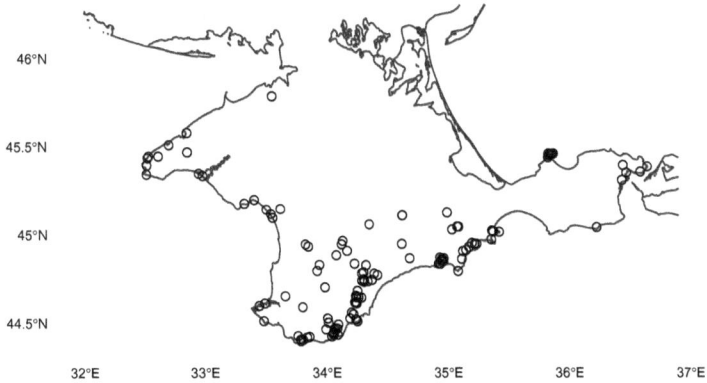

Fig. 1. Actual localities of *P. tragium* on the Crimean Peninsula.

The work reported was conducted completely in R free statistical software which enabled building script for reproducible research without using any additional software [9].

4   Results and Discussion

4.1   Data Preparation and Optimization

The initial dataset on bioclimatic variables comprised 19 variables of which 11 characterizes temperature regime and 8 quantify precipitation (see Table 1). The bioclimatic variables represent annual trends (e.g., mean annual temperature, annual precipitation), seasonality (e.g., annual range in temperature and precipitation), and extreme or limiting environmental factors (e.g., temperature of the coldest and warmest month, and precipitation of the wet and dry quarters) [6].

Four variables combine two types of the climate data being dependant on both precipitation and temperature (BIO 8, BIO 9, BIO 18, and BIO 19). Among the named two groups BIO 2, BIO 3, BIO 4, and BIO 7 are derived from others arithmetically and thus are to be correlated. Pairs of variables like BIO 5 and BIO 10 or BIO 13 and BIO 16 also appear to be in a correlation. Using the complete set of the variables as predictors for SDM training would lower model performance due to noise introduced with the voluminous data. This poses the need in optimization of input data for SDMs - whether the data are bioclimatic or of some other nature. In order to optimize the data we conducted exploratory analysis of predictor dataset by means of unsupervised machine learning techniques. First, to unravel the data multivariate structure hierarchical clusterization using Ward's method was performed and its results were assessed in a form of dendrogram (Fig. 2). This highlighted the presence of 5 distinct clusters which were also plotted in the geographical map (Fig. 3); the latter were mostly in agreement with commonly recognized climatic regions of the Crimea [9]. Second, correlation analysis using Pearson coefficient was made (Fig. 4). As expected, many

Table 1. Bioclimatic variables.

Code	Variable
BIO 1	Annual mean temperature
BIO 2	Mean diurnal range (mean of monthly (max temp − min temp)
BIO 3	Isothermality (BIO2/BIO7) (* 100)
BIO 4	Temperature seasonality (standard deviation * 100)
BIO 5	Max temperature of warmest month
BIO 6	Min temperature of coldest month
BIO 7	Temperature annual range (BIO5–BIO6)
BIO 8	Mean temperature of wettest quarter
BIO 9	Mean temperature of driest quarter
BIO 10	Mean temperature of warmest quarter
BIO 11	Mean temperature of coldest quarter
BIO 12	Annual precipitation
BIO 13	Precipitation of wettest month
BIO 14	Precipitation of driest month
BIO 15	Precipitation seasonality (coefficient of variation)
BIO 16	Precipitation of wettest quarter
BIO 17	Precipitation of driest quarter
BIO 18	Precipitation of warmest quarter
BIO 19	Precipitation of coldest quarter

variables were found to be well correlated. This was especially substantial for the precipitation data. Later on principal component analysis (PCA) was carried out and presented as biplot in a space of two first principal components (Fig. 5). Considering two first PCs seemed to be sufficient since they accounted for over 70% of the dataset variance, as evidenced by scree plot (not shown). The PCA results also suggest high correlation among the variables: precipitation variables shown on the plot as arrows have similar direction and tend to form a bundle. This implies their rather uniform contribution to the variance of the data and that each separate variable holds little unique (being absent in others) information. On the contrary, directions of the temperature variables are diverse, yet some of them still are clumped. Their overall impact on variance however exceeded one of precipitation data. In addition to direction, the representing initial variables arrows are also characterized by length which represents the magnitude of their impact on variance. The biplot imply substantial importance of multiple variables, including all the variables from the precipitation group and little importance of some, mainly derived from the others from the data (BIO 2, BIO 3, BIO 9). As a result of this stage of study, the list of most correlated and less informative variables was established as follows: BIO 3, BIO 12, BIO 13, BIO 14, BIO 15, BIO 17. The variables were excluded from the initial dataset and absent in the new reduced one.

The second round of exploratory analysis using the aforementioned techniques was applied to the reduced dataset with subsequent comparison to the initial previously described. Hierarchical clusterization output in both cases were compared in a form of

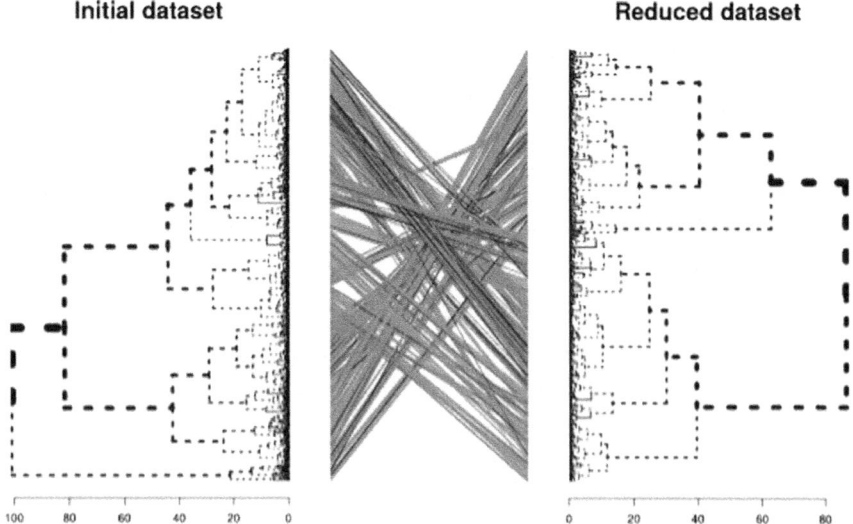

Fig. 2. Tanglegram comparing dendrograms representing hierarchical clusterization for initial and reduced dataset.

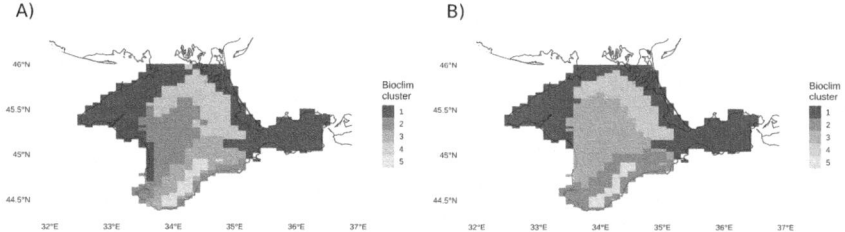

Fig. 3. Hierarchical clusterization results plotted on map for: A. initial dataset and B. reduced dataset.

dendrograms (Fig. 2). The optimal number of clusters for the dendrograms was equal to 5. Alignment of the two dendrograms ('tanglegram') with identical nodes joined by straight lines is shown on Fig. 2. Here matching subtrees are joined by grey lines; the rest of the lines are shown in black. Overall two tanglegrams share high similarity, particularly on the level of smaller branches. The similarity described was additionally quantified by means of cophenetic correlation coefficient, which established to be 0.77. Pearson's correlation matrices showed decreased number of correlated variables in the case of new dataset. PCA evidenced that the arrows corresponding to variables became more sparse and less prone to join into bundles. One can therefore suggest that the data multivariate structure in general is preserved through the reduction step. In conclusion, the data optimization step yielded more compact dataset with reduced dimensionality with most of the initial information sustained.

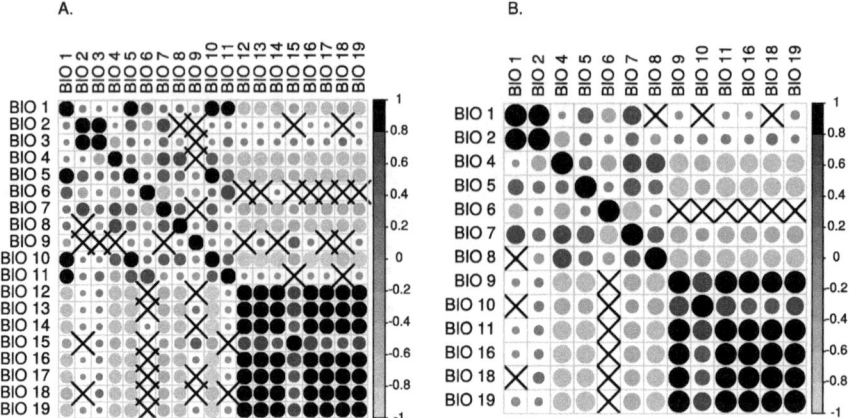

Fig. 4. Correlation coefficient visualization for: A. initial dataset and B. reduced dataset. Insignificant values (cutoff P = 0.01) are crossed out. Abbreviation see Table 1.

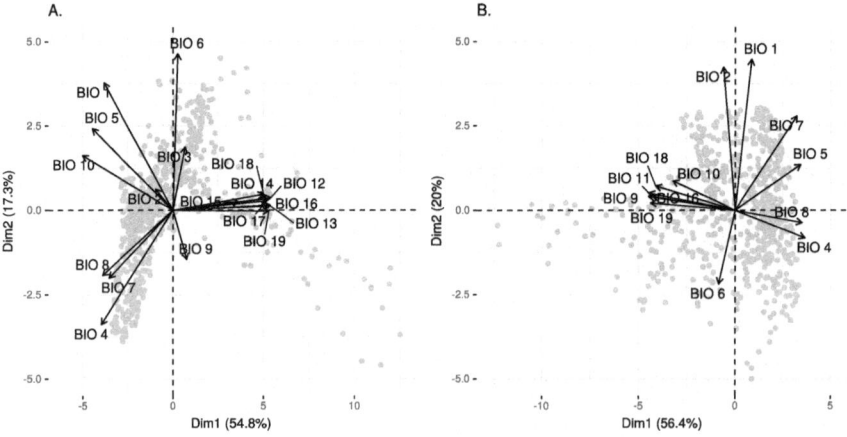

Fig. 5. Biplot representing PCA analysis for: A. initial dataset and B. reduced dataset, principal components 1 and 2 space. Points represent sites, arrows represent variables. Abbreviation see Table 1.

4.2 Spatial Distribution Models

Both initial and reduced dataset were used for SDMs training and comparison. Actual localities of *P. tragium* came from herbarium specimen collected across the peninsula of Crimea. The complete set of coordinates for georeferencing points (131) were split into training and testing sets (79 and 52 points, accordingly). The background data were derived for the rectangular grid for the whole peninsula. Randomly pooled from the sets of 130 and 195 points were joined with the actual localities and after the scaling procedure served as the set of predictors for binary classifiers training and evaluation. Since Maxent is the algorithm that is used for the task widely but not exclusively, we compared the method (implemented in R statistical software [9]) with various other techniques accessed through caret R package (namely, plsRglm: Partial Least Squares Regression for Generalized Linear Models, LogitBoost, gbm: Generalized Boosted Regression Models, mlpML: Multilayer Perceptron, nb: Naive Bayes, rf: Random Forest, svmRadial: Support Vector Machines with Radial Basis Function Kernel). As the input data both the intact set of 19 variables and the reduced one in which 6 variables were removed (namely, BIO 3, BIO 12, BIO 13, BIO 14, BIO 15, BIO 17). The machine learning results were visualised in a form of map showing predicted distribution areas in each case as well as actual localities sites. The predicted distributions showed good agreement in most cases. They are mostly follow the pattern apparent from visualizing actual localities in the same map. *P. tragium* therefore is expected to grow primarily at the southern coast (Fig. 6). Next the performance of models obtained was assessed by their accuracy, sensitivity, and specificity. Overall the models performed very well, except for plsRglm having low performance and somehow different predicted distribution areas. This however might be connected to its technical peculiarity and not valid performance. The input data reduction procedure slightly increased performance values in some cases and tipped the sensitivity-specificity trade off in others (Fig. 7). The improvement is far dramatical, yet it was obtained with the initial data significantly reduced in size.

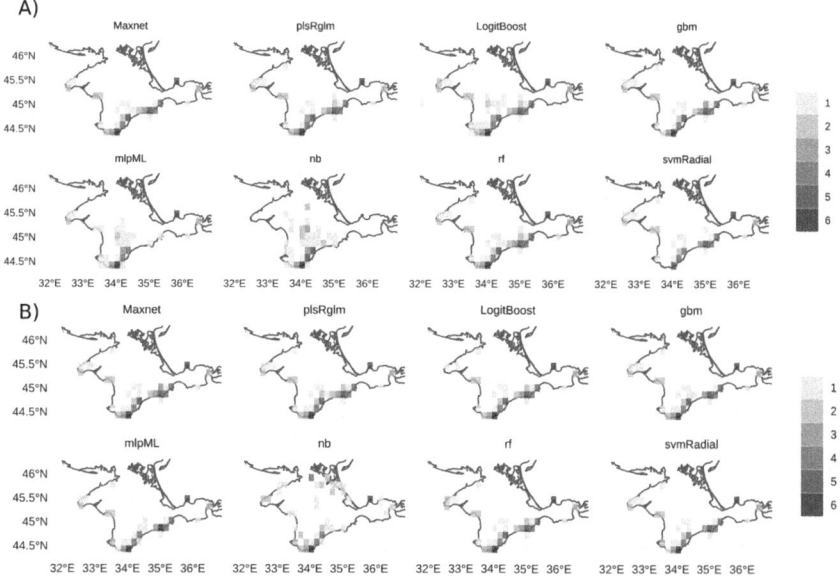

Fig. 6. Predicted distribution areas for *P. tragium* by models trained using: A. the initial datset and B. the reduced dataset.

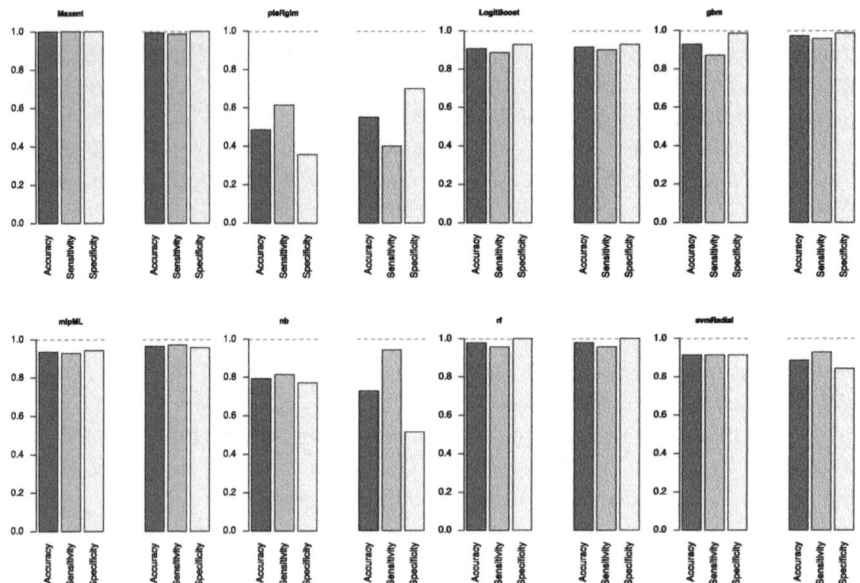

Fig. 7. Performance measures for the model trained using the initial dataset (left Accuracy, Sensitivity, and Specificity bars on each of paired barplots) and reduced dataset (rights).

5 Conclusion

The work primarily aimed at exploring the possibility of input data preparation and optimization for the advancement of spatial distribution models (SDMs). The need stems from the notion that using highly correlated and less impactful variables is known to lower performance of machine learning techniques. This task was fulfilled for the number of model types, including Maxnet. Besides, analogous data reduction could help in quantifying the impact of a variable on the classifiers work. In case of Maxent algorithm the step is usually performed using permutation test. For the test some predictors are modified randomly with the consequent evaluation of performance alteration in models trained. Here we reported approach comprising data processing prior to machine learning step which is purely data driven and does not depend on a classifier training. The data optimization encompassed clusterization, correlation assessment, and principal component analysis (PCA) with the purpose to establish variables that are most contributing and hence impactful. Using the workflow, one can additionally perform exploratory data analysis in order to elucidate multivariate patterns in the dataset. This enables familiarizing a practitioner with the data structure and might help in analysis with the aim of research. Notably, such investigation important due to the fact that importance of a variable in each particular prediction may depend not only on the parameter nature and interconnection to other variables, but also on case region. Indeed, territories located in remote regions may have different contribution of the same environment characteristic. In the current study exploratory merit of the approach for bioclimatic variables set of the Crimean peninsula is that precipitation variables are well correlater and affect variance uniformly, yet magnitudes of their impact are comparable. Applying predictors assessment before classifiers training also allows to use the knowledge obtained for building classifier that are not Maxent-based. One can speculate that augmenting Maxent usage with some other machine learning algorithms could improve robustness of the resulting predictions and provides researchers with wider tool spectrum. In the reported study comparison of Maxent against several algorithms representing fairly different groups has confirmed the suggestion.

References

1. Franklin, J: Mapping Species Distributions: Spatial Inference and Prediction. Cambridge University Press, Cambridge, UK. ISBN 9780521876353 (hbk), vol. **44**, p. 615. (2009) https://doi.org/10.1017/s0030605310001201
2. Elith, J, Leathwick, J.R.: Species distribution models: ecological explanation and prediction across space and time. Ann. Rev. Ecol. Evol. Syst. **40**, 677–697 (2009) https://doi.org/10.1146/annurev.ecolsys.110308.120159
3. Phillips, S.J., Dudík, M.: Modeling of species distributions with Maxent: new extensions and a comprehensive evaluation. Ecography (2008). https://doi.org/10.1111/j.0906-7590.2008.5203.x
4. Elith, J. et al.: Novel methods improve prediction of species' distributions from occurrence data. J. Space Time Ecol. (2006). https://doi.org/10.1111/j.2006.0906-7590.04596.x
5. Ni, L., Cook, D., Tsai, C.-L.: A note on shrinkage sliced inverse regression. Biometrika **92** (1), 242–247 (2005). https://doi.org/10.1093/biomet/92.1.242
6. Fick, S.E., Hijmans, R.J.: Worldclim 2: new 1-km spatial resolution climate surfaces for global land areas. Int. J. Climatol. (2017). https://doi.org/10.1002/joc.5086
7. Elith, J. et al.: A statistical explanation of MaxEnt for ecologists. Divers. Distrib. **17**, 43–57 (2011). https://doi.org/10.1111/j.1472-4642.2010.00725.x
8. Shipley, B.: Inferential permutation tests for maximum entropy models in ecology. Ecology. (2010). https://doi.org/10.1890/09-1255.1
9. R Statistical Programming Environment: https://www.r-project.org/
10. Klimaticheskiy atlas Kryma: prilozhenie k nauchno-prakticheskomu diskussionno-analiticheskomu sborniku «Voprosy razvitiya Kryma» [Climatic Atlas of the Crimea: an Application to Scientific and Practical Discussion-Analytical Collection «The Development of Crimea»]. Simferopol, 120 p. (2000)

Forest Resources of the Baikal Region: Vegetation Dynamics Under Anthropogenic Use

Anastasia K. Popova[1(✉)] ⓘ, Evgeny A. Cherkasin[1],
and Igor N. Vladimirov[2]

[1] Matrosov Institute for System Dynamics and Control Theory SB RAS, Irkutsk,
Russia
chudnenko@icc.ru
[2] V.B, Sochava Institute of Geography SB RAS, Irkutsk, Russia

Abstract. We review the theoretical and applications-specific issues of modeling a temporal and spatial dynamics of forest ecosystems, based on the principles of investigating dynamical models. The model used takes into account various factors affecting the change in forest areas–fires, forest diseases, cutting, urban expansion, etc. Calculation of numerous scenarios for the use of forest resources allows us to see the consequences of various managerial decisions. The user can adjust the parameters of the main cutting volume, fires and tree planting. We present the results of a computer modeling and predictive mapping for the regional model of the forest resource dynamics under anthropogenic use.

Keywords: Forest resource dynamic · Forest map · Forest modelling

1 Introduction

Ecological forecasting and modeling are important tools in studying the dynamics of forest resources. A correct evaluation of the parameters of vegetation change makes it possible to build models that are closest to reality. Information systems which interact with forest management modeling systems and take into account anthropogenic use provide management decision making. Calculating effects of implementing various managerial decision helps to evaluate the development of forests, depending on the conditions of each scenario.

There are a number of papers devoted to the study of the influence of anthropogenic factors on the dynamics of forest resources. Mladenoff and Scheller used models to evaluate the complex effects of climate change, harvest, wind, species migration on the dynamics of regional forests in northern Wisconsin [1]. Abood, Lee et al. investigated the contribution of logging and mining, palm oil harvesting on forest losses in Indonesia [2]. Popradit et al. analyzed the effects of settlement expansion on area and species diversity of tropical forest trees [3]. Musi et al. studied the impact of agricultural development on forest land reduction on the example of Central Java [4]. Wu et al. integrated land use change models based on CA-Markov and forest landscape

© Springer Nature Switzerland AG 2019
I. Bychkov and V. Voronin (Eds.): *Information Technologies
in the Research of Biodiversity,* SPEES, pp. 96–106, 2019.
https://doi.org/10.1007/978-3-030-11720-7_14

model LANDIS-II to simulate dynamic of the forest landscape in response to the disturbances of land use change and harvest in the Taihe district, China [5].

This study is a continuation of the work done by authors [6–8]. We have created a software tool designed to automate decision support for the use of forest resources in the Baikal region of the Russian Federation. The developed program consists of a set of subsystems, including a geoinformation system (GIS) used to display spatially-distributed information, a mathematical modeling block and an interface with the function for forming various scenarios of forest dynamics. In the work, Model "Dynamics of stands" was used, which has been designed to calculate by age classes over vast areas, taking into account economic development of the territory.

In this work, an analysis of various scenarios for the quantitative assessment of the influence of a number of factors on the change of forest areas in the Baikal region has been carried out. The objectives of the study are to: (1) modeling the dependence of the dynamics of forest resources on different types of anthropogenic factors; and (2) quantitative assessment of the impact of changes in anthropogenic factors on forest dynamics. The quantitative assessment is designed to make a regional strategy for the use of forest resources.

2 Materials and Methods

2.1 Study Area

This study was carried out in Irkutsk region of the Russian Federation. The region is located in the center of Eurasia, in Eastern Siberia at coordinates $51° 18'$–$64° 15'$N, $95° 38'$–$119° 10'$E and occupies 774846 km^2. Two thirds of the area is covered with forests. Pale conifers predominate - pine and larch. Also in the territory grow dark coniferous - spruce, fir and cedar - and deciduous trees - birch, aspen.

2.2 Mathematical Model

The model "Dynamics of stands" is based on the works of Cherkashin [9–11], taking into account the studies of Shifley, He [12], Shugart [13, 14], Wu [15], Horn [16], Mladenoff [17]. The dynamics is described by a system of differential equations. Land areas of different types are made on the modeled territory: non-forested, uncovered by forest, young, middle-aged, maturing, mature and over-mature. Non-forest area is an area where forests cannot grow. It includes settlement, road, and deposit areas. The surface of non-forested areas is not covered by trees temporarily. It's an area after the fire, cutting, damage by insects and weather conditions. The dynamics of each section is described by formulas as follows (1):

$$\frac{dS_N}{dt} = -a_{N0}S_N(t) + u_{nonN}(t),$$

$$\frac{dS_0}{dt} = a_{N0}S_N(t) - a_{01}S_0(t) + u_{ncov0}(t) + u_{cut0}(t) - u_{non0}(t), \tag{1}$$

$$\frac{dS_i}{dt} = a_{i-1i}S_{i-1}(t) - a_{ii+1}S_i(t) - u_{noni}(t) - u_{ncovi}(t) - u_{cuti}(t),$$

where a_{ij} are the coefficients of transition from one category of land or age group to the next;

S_N is the non-forest area;

S_0 is an area that is uncovered by forest;

S_i is forest areas of different classes of age;

$u_{non\ i}$ is annual increase in non-forest area at the expense of other categories of land;

$u_{ncov\ i}$ is increase in the area uncovered by forest;

$u_{cut\ i}$ is the area of cutting, is subtracted only from the category of mature and over-mature forests, in other classes of age cutting is not carried out.

The annual decrease of the volume of forest resources is due to the impact of natural adverse factors and anthropogenic use. There are a number of permitted uses for forest resources of the Irkutsk region:

- cutting;
- agriculture;
- recreational activity;
- development of mineral deposits, work on geological study of mineral resources;
- construction and operation of linear objects;
- building and operation of reservoirs and other hydraulic structures;
- processing of wood and other forest resources.

Harvesting of wood is carried out in operational forests. It includes clear and selective cutting of mature and overmature plantations, clear and selective sanitary cutting, care, and other cutting. The annual allowable cut for forestry in the Baikal region is 71.5 million m^3. At the same time, the actual volumes of cutting allow us to reach the estimates only by 40%.

The area of forest areas used for cultural and recreational purposes is increasing for the development of tourism. Recreational activity is being carried on in the Baikal region with access to Lake Baikal, infrastructure and unique natural objects are being developed on its territory.

The exploitation areas of forests are involved in the mining and subsoil exploration, and a part of the reserve forests is transferred to operational ones. Construction and maintenance of hydraulic structures is carried out on the territory of that forestry, where modern wood processing enterprises are being created.

Therefore, the increase in non-forest area in the process of forest exploitation is as follows:

$$u_{non} = k_N \Delta N + \Delta S + \Delta R + \Delta G + \Delta Bl + \Delta Bv, \tag{2}$$

where k_N is the area of settlements per person, the remaining coefficients characterize the increase of forest population, ΔN, agricultural area, ΔS, recreational zones, ΔR, area of fields, ΔG, construction of linear objects, ΔBl, and maintenance of hydraulic structures, ΔBv.

The transition of other categories of land to non-forest ones is done at random, depending on the needs of production. On this basis, the distribution of u_{non} for the remaining categories of land is assumed to be proportional to the current area of each region:

$$u_{non\,i} = u_{non} * \frac{S_i}{\sum_i S_i} \tag{3}$$

Summation is made for all breeds and classes of age, including uncovered forest areas.

Annually forests are exposed to a set of adverse factors. The weakening and destruction of forest plantations in the Irkutsk region is affected by:

- fires;
- forest diseases;
- damage by insects;
- adverse weather conditions;
- other anthropogenic and non-pathogenic factors.

The fires (43.7%), forest diseases (25.2%) and insect damage (16.2%) are the most important factors for drying up and destruction of forest resources, so these factors should be taken into account in the model:

$$u_{ncov} = S_g + S_{nas} + S_b \tag{4}$$

The distribution of u_{ncov} by categories of land considered to be similar to the previous one:

$$u_{ncov\,i} = u_{ncov} * \frac{S_i}{\sum_i S_i} \tag{5}$$

The pyrogenic factor is the main reason for the weakening of plantations and death of forests. The calculations take into account only the area of forest resources with the lost stability, not including all the lands passed by fires.

Damage by insects causes shrinking and weakening of trees, a decrease in growth, causing unsatisfactory condition of forest resources. The calculations also take into account only plantations with lost resistance, not including all disturbances.

2.3 Software

The complex software system is implemented in the Java programming language. The calculation block loads the initial data for modeling from Microsoft Excel tables. They provide information on the forestry: the distribution of areas by land categories and classes of age, the volume of cutting, stocks, forest plantation areas, the number of people living, the values of the parameters of anthropogenic use and natural factors.

The indicator of the population increases every year by a certain amount of annual population growth defined in a model scenario. The values of the parameters of anthropogenic use and natural factors for each forest area are considered constant throughout the simulation period.

After starting the program, the user sets the input values – a forecasting interval and the step size. A numerical solution of the system of differential equations is performed by Runge-Kutta method of the fourth order. The combination of parameters – the volume of cutting, forest plantations and the impact of adverse conditions – is set as a percentage of the currently available values and forms a resource management scenario (Fig. 1).

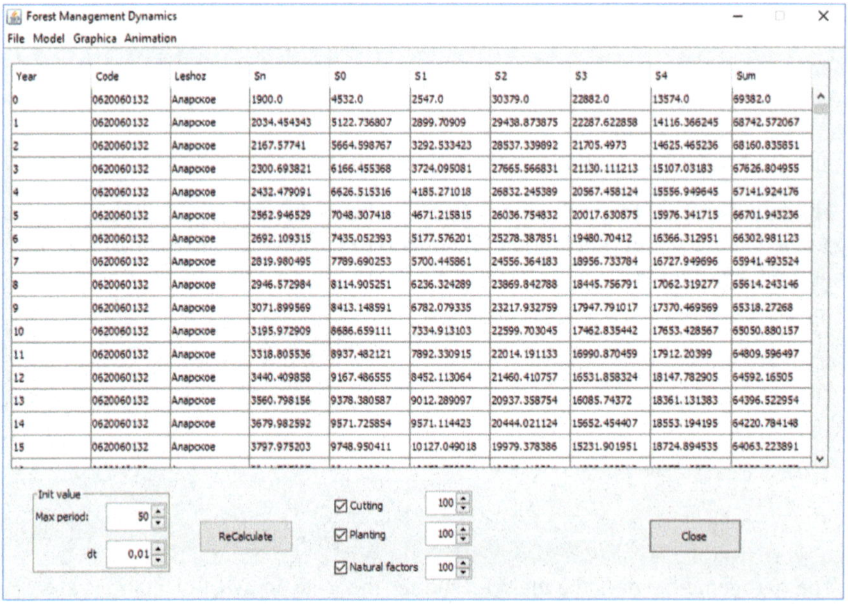

Year	Code	Leshoz	Sn	S0	S1	S2	S3	S4	Sum
0	0620060132	Анарское	1900.0	4532.0	2547.0	30379.0	22882.0	13574.0	69382.0
1	0620060132	Анарское	2034.454343	5122.736807	2899.70909	29438.873875	22287.622858	14116.366245	68742.572067
2	0620060132	Анарское	2167.57741	5664.598767	3292.533423	28537.339892	21705.4973	14625.465236	68160.835851
3	0620060132	Анарское	2300.693821	6166.455368	3724.095081	27665.566831	21130.111213	15107.03183	67626.804955
4	0620060132	Анарское	2432.479091	6626.515316	4185.271018	26832.245389	20567.458124	15556.949645	67141.924176
5	0620060132	Анарское	2562.946529	7048.307418	4671.215815	26036.754832	20017.630875	15976.341715	66701.943236
6	0620060132	Анарское	2692.109315	7435.052393	5177.576201	25278.387851	19480.70412	16366.312951	66302.981123
7	0620060132	Анарское	2819.980495	7789.690253	5700.445861	24556.364183	18956.733784	16727.949696	65941.493524
8	0620060132	Анарское	2946.572984	8114.905251	6236.324289	23869.842788	18445.756791	17062.319277	65614.243146
9	0620060132	Анарское	3071.899569	8413.148591	6782.079335	23217.932759	17947.791017	17370.469569	65318.27268
10	0620060132	Анарское	3195.972909	8686.659111	7334.913103	22599.703045	17462.835442	17653.428567	65050.880157
11	0620060132	Анарское	3318.805536	8937.482121	7892.330915	22014.191133	16990.870459	17912.20399	64809.596497
12	0620060132	Анарское	3440.409858	9167.486555	8452.113064	21460.410757	16531.858324	18147.782905	64592.16505
13	0620060132	Анарское	3560.798156	9378.380587	9012.289097	20937.358754	16085.74372	18361.131383	64396.522954
14	0620060132	Анарское	3679.982592	9571.725854	9571.114423	20444.021124	15652.454407	18553.194195	64220.784148
15	0620060132	Анарское	3797.975203	9748.950411	10127.049018	19979.378386	15231.901951	18724.894535	64063.223891

Fig. 1. Interface of program.

The output of the simulation is presented in a tabular form for the forest areas, years and age categories. The resulting tables are saved to a CSV file by user request. Freely distributed JFreeChart library is used for the graph construction. It displays curves of area changes for each age category with a different color on the same figure. To obtain the totals, the indicators of the same type are preliminarily summed up for all the forest districts, giving the final value throughout the Baikal region.

The GIS subsystem builds maps based on SHP cartographic material using free OpenMap library. Each forest area is painted with a certain color, depending on the size of the selected parameter. In the resulting coloring, a lighter color corresponds to a smaller value of the parameter, a darker one to a larger value.

To create a map, the user selects the category of lands to display, the estimated data for this category is divided into five groups, and each forestry department is assigned the corresponding coloring value as a result. Relative values are taken to determine ranges: the ratio of the difference in the current area of the selected category of land and its initial value to the total area of the forest area. The finished map is saved as a JPEG image by user request.

3 Results and Discussion

Six different scenarios were computed to estimate the dynamics of the forest resources of Irkutsk region for 50 year interval. Forecasts were based on the data for 2017 by categories of land and age classes of forest resources of all 37 forestries in the territory of the Baikal region. Units of measure are total areas in thousand hectares in all forest districts.

Six scenarios are considered:

1. Current levels of cutting, tree planting and the influence of adverse factors.
2. Reducing the influence of adverse factors by 2 times.
3. Increase in the volume of tree planting by 2.5 times.
4. Increase of cuttings by 3 times.
5. Increase cuttings by 2.5 times, planting by 1.5 times.
6. Decrease of cuttings by 2 times.

The simulation results are shown on the map of the Irkutsk region, as a measure taken the difference between the forest area for all categories of age at the end and the beginning of the simulation interval. This approach helps to quickly assess the dynamics of the resource - with a decrease in forest areas over a period, we get negative values of the parameter, with an increase - positive. In Fig. 2, colors 1–2 correspond to negative values relative to the initial values, it is a decrease, color 3 is minor fluctuations, colors 4–5 is positive values, i.e. their area increases.

The results show that the increase in the level of cuttings in scenario 4 causes a significant decrease in forest areas in the central part of the region. Reduction of adverse factors in Scenario 2, increase in planting in Scenario 3, and reduction of cutting in the latter variant leads to an increase in the number of forest lands. In the combined scenario 5, the increase in cutting is not sufficiently compensated by the increase in forest plantations–the forest areas in the 50-year estimated period are markedly reduced.

It can be seen that forest areas located in the central part of the region are most affected by anthropogenic influence. In large forest areas in the north and east of the region, due to low transport accessibility, forest areas remain almost at the same level, regardless of the scenario of their use.

We make a consolidated quantitative assessment of the change in the total forest area in the scenario simulations. The results of the comparison are given in Table 1. For each scenario, the total forest area at the end of the simulation period was calculated (the rows "Area") and its difference with scenario 1, taking into account only the current exposure levels, in% (the rows "Diff").

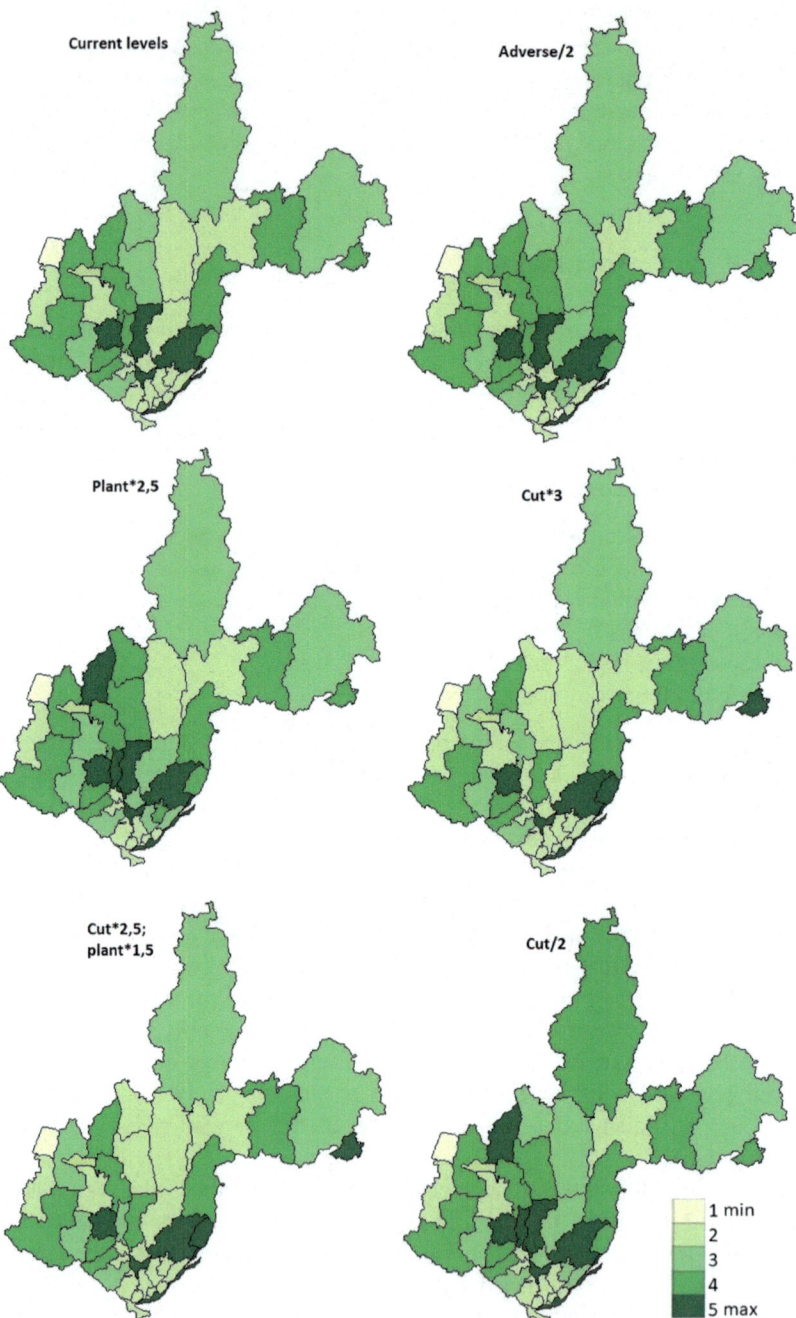

Fig. 2. Simulated land use maps generated under six scenarios.

If the influence of adverse factors is reduced by 2, the total area will increase by 0.4% (236 thousand hectares), as compared to the calculation at current levels of factors. With an increase in forest plantations in 2.5 times, an increase is noted of the total area by 0.5% (320 thousand ha).

Increasing the cutting by 3 times, results in a decreasing of the total area by 1.3% (780 thousand hectares), with the areas of mature and over-mature trees decreasing by 26.6%, while the young forests area is increased by 31.3. In scenario with an increase in logging by 2.5 times (which corresponds to the full development of the estimated cutting area) with a simultaneous increase in forest plantations in 1.5 times, a decrease in total area by 0.9% (568 thousand hectares) is observed. %. If the volume of cuttings is reduced by 2 times, the total area will increase by 0.8% (501 thousand ha), while mature and over-mature area grows by 11.5%.

Table 1. Quantitative changes in areas in the computer simulations of defined scenarios with respect to the current level of factors.

Scenario		Current	Adverse/2	Plant*2.5	Cut*3	Cut*2.5; plant*1.5	Cut/2
Young	Area	12467	11733	12674	16367	15730	10783
	Ratio		−5.9	1.7	31.3	26.2	−13.5
Middle-aged	Area	13968	13998	14054	15976	15608	13217
	Ratio		0.2	0.6	14.4	11.7	−5.4
Maturing	Area	6755	6900	6773	7225	7134	6592
	Ratio		2.2	0.3	6.9	5.6	−2.4
Mature and over-mature	Area	26932	27727	26940	19774	21083	30031
	Ratio		2.9	0.02	−26.6	−21.7	11.5
Total	Area	60122	60358	60442	59342	59554	60623
	Diff		236	320	−780	−568	501
	Ratio		0.4	0.5	−1.3	−0.9	0.8

The graph in Fig. 3 shows that the decrease in adverse factors (43.7% of them are fires) in Scenario 2 over the first 40 years will give a larger increase in the total area than the increase in forest plantations by 2.5 times in Scenario 3. Only after 40 years increase in plantings will begin to give a slightly larger increase in area. The sharp decrease in the area due to the increase in cuttings in 3 times in Scenario 4 is compensated by an increase in forest plantations by 1.5 times with an increase in cutting by 2.5 times in Scenario 5.

The results of the simulation showed that an increase in the volume of cutting for the development of the estimated cutting area will lead to a significant reduction of forest areas. If, however, the increase in cuttings is supplemented by an increase in planting of trees and intensification of the fight against adverse factors, especially forest fires and tree diseases, then it will be much sooner to compensate for the reduction of forests.

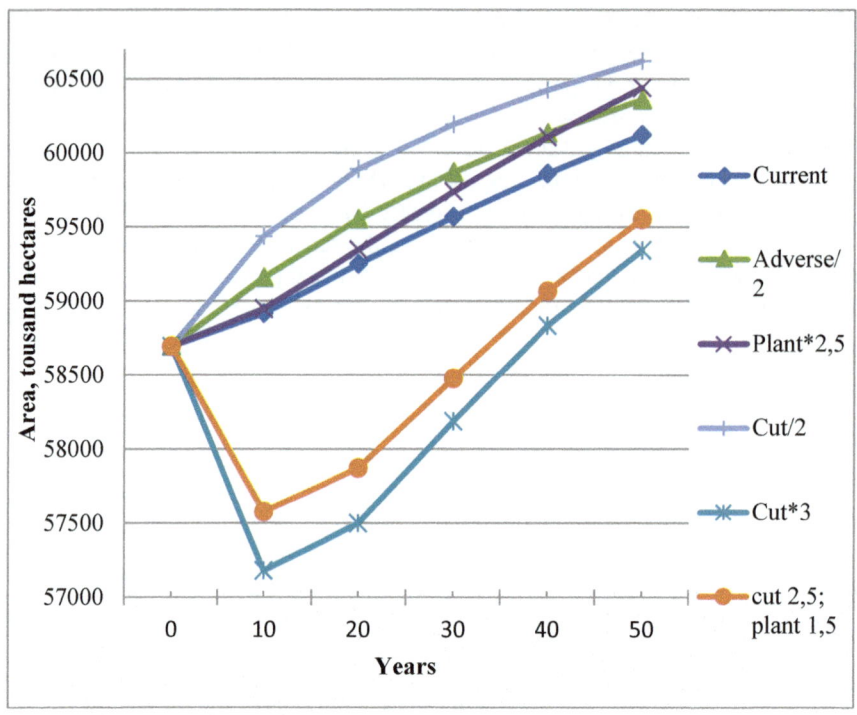

Fig. 3. Changes in total forest area according to 6 scenarios for 50 years.

4   Conclusions

The developed software can simulate the dynamics of forest resources taking into account the influence of a complex of natural and anthropogenic factors. It also helps to make managerial decisions in the forestry sector, showing the direction of changing the area of a category of land depending on the given scenario.

The results of calculations are presented to the user in the form of tables, graphs and maps. The final values are displayed for each year from the given time interval for each category of land and the age class of the trees. One can analyze the results of modeling with these tools.

When calculating forecast scenarios, a strategy is found that ensures a balance between the maximum volume of harvesting and the preservation of forest areas significant reduction: an increase in cutting by a 2.5 times will not lead to a sharp depletion of reserves with a simultaneous increase in forest plantations by a 1.5 times. In this case, the areas of mature and over-mature after a small decrease during the first 10 years will later be restored. Reduction of adverse factors in 2 times causes an increase in forest areas by 0.4%, comparable to an increase in plantings by 2.5 times - 0.5%.

However, not all possible natural and anthropogenic factors affecting the changes in the forest landscape are taken into account in scenario calculations. To obtain more accurate results, it is necessary to take into account other factors: climate, harvesting, urban planning. Also, the calculations are based on the assumption that the rates of transition from one category of land to another and the values of the considered factors in the future will match with those currently available. Such assumptions reduce the accuracy of the results, but allow us to simplify the modeling process.

The combination of methods of mathematical modeling with the features of geographical mapping provides integration of information flows in forestry activities, visualization of information about the spatial and temporal state of forest resources. The results are to be used to support decision-making in the field of forest regional management, for achieving the economic development and the preservation of an environment comfortable for human beings.

Acknowledgements. The work was carried out with financial support: Russian Foundation for Basic Research grant №: 18-07-00758-a; Integration program ISC SB RAS. Results are achieved using the Centre of collective usage «Integrated information network of Irkutsk scientific educational complex».

References

1. Scheller, R.M., Mladenoff, D.J.: A spatially interactive simulation of climate change, harvesting, wind, and tree species migration and projected changes to forest composition and biomass in northern Wisconsin, USA. Glob. Chang. Biol. **11**, 307–321 (2005)
2. Abood, S.A., Lee, J.S.H., Burivalova, Z., Garcia-Ullo, A.J., Koh, L.P.: Relative contributions of the Logging, Fiber, Oil Palm, and mining industries to forest loss in Indonesia. Conserv. Lett. **8**, 58–67 (2015)
3. Popradit, A., Srisatit, T., Kiratiprayoon, S., Yoshimura, J., Ishida, A., et al.: Anthropogenic effects on a tropical forest according to the distance from human settlements. Sci. Rep. **5**, 14689 (2015)
4. Musi, C., Anggoro, S., Sunarsih: System dynamic modelling and simulation for cultivation of forest land: case study Perum Perhutani, Central Java, Indonesia. J. Ecol. Eng. **18**(4), 25–34 (2017)
5. Wu, Z., Ge, Q., Dai, E.: Modeling the relative contributions of land use change and harvest to forest landscape change in the Taihe County, China. Sustainability **9**, 708 (2017)
6. Popova, A.K., Vladimirov, I.N.: Multilevel Modeling of the Forest Resource Dynamics. Mathematical Modelling of Natural Phenomena, vol. 4(5), pp. 72–88 (2009)
7. Vladimirov, I.N., Chudnenko, A.K.: Forecasting of the spatio-temporal dynamics of the forest resources of the Irkutsk region using GIS technologies. Sun Earth, Water, and Energy. Transactions of the Siberian Division of APVN, issue 2, pp. 61–68. Nauka, Novosibirsk (2005)
8. Cherkashin, E.A., Badmatsyrenova, S.B., Vladimirov, I.N., Popova, A.K., Davydov, A.: An optimal control module of sustainable natural resources consumption control synthesis for decision support systems. In: 37th International Convention on Information and Communication Technology, Electronics and Microelectronics (MIPRO), pp. 1100–1105 (2014)
9. Cherkashin, A.K.: Forecasting the spatial and temporal dynamics of forests of taiga landscape. Dynamics of Ecologo-Economic Systems, pp. 107–111. Nauka, Novosibirsk (1981)

10. Cherkashin, A.K.: The model for the dynamics of forest stands of a district forestry and its use in solving prediction problems. Planning and Forecasting of Natural-Economic Systems, pp. 69–81. Nauka, Novosibirsk (1984)
11. Cherkashin, A.K.: The expanding complex of particular models. Forest Systemic Research into a Region's Nature–Economy Interaction, pp. 71–77. Irkutsk University Publisher (1986)
12. Shifley, S.R., He, H.S., Lischke, Y., Wang, W.J., et al.: The past and future of modeling forest dynamics: from growth and yield curves to forest landscape models. Landscape Ecol. **32**, 1307–1325 (2017)
13. Armstrong, A., Fischer, R., Shugart, H., Huth, A., Fatoyinbo, L.: Simulating forest dynamics of lowland rainforests in Eastern Madagascar. Forests **9**(4), 214 (2018)
14. Shugart, H.H.: A Theory of forest dynamics. The Ecological Implications of Forest Succession Models. Springer, N.Y (1984)
15. Wu, J., David, J.L.: A spatially explicit hierarchical approach to modeling complex ecological systems: theory and applications. Ecol. Model. **153**, 7–26 (2002)
16. Horn, H.S.: Some causes of variety in patterns of forest succession. Forest Succession: Concepts and Applications, pp. 24–35. Springer, N.Y. (1991)
17. Mladenoff, D.J., He, H.S.: Design and behavior of LANDIS, an object-oriented model of forest landscape disturbance and succession. Advances in Spatial Modeling of Forest Landscape Change: Approaches and Applications, pp. 125–162. Cambridge University Press (1999)

DNA Barcoding of *Waldsteinia* Willd. (Rosaceae) Species Based on ITS and *trnH-psbA* Nucleotide Sequences

Marina Protopopova[1](✉) (iD), Vasiliy Pavlichenko[1] (iD),
Aleksander Gnutikov[2,3] (iD), and Victor Chepinoga[4,5] (iD)

[1] Siberian Institute of Plant Physiology and Biochemistry SB RAS, Lermontov Street 132, 664033 Irkutsk, Russia
{marina.v.protopopova, vpavlichenko}@gmail.com
[2] The N.I. Vavilov Institute of Plant Genetic Resources (VIR), Bolshaya Morskaya Street 44, 190000 St.-Petersburg, Russia
alexandr2911@yandex.ru
[3] Komarov Botanical Institute of the Russian Academy of Sciences, Professora Popova Street 2, 197376 St.-Petersburg, Russia
[4] The V.B Sochava Institute of Geography SB RAS, Ulan-Batorskaya Street 1, 664033 Irkutsk, Russia
Victor.Chepinoga@gmail.com
[5] Irkutsk State University, Karl Marks Street 1, 664003 Irkutsk, Russia

Abstract. The methods of biological species identification using nucleotide sequences of short genome regions (DNA markers) are actively developed. This principle formed the basis of the genetic database purposed the identification for all living organisms—Barcode of Life Data System (BOLD). In our study, we estimated the Internal transcribed spacer region (ITS) and *trnH-psbA* intergenic chloroplast spacer as possible markers for species identification of vascular plant genus *Waldsteinia* (Rosaceae). The difficulties in morphological distinguishing of several *Waldsteinia* species require the development of approaches using DNA barcodes for the species identification. The comparative analysis of the nucleotide sequences of East Asian *Waldsteinia maximowiziiana*, South Siberian *W. ternata* (from the Khamar-Daban Ridge and the Eastern Sayan Mts.) and *W. tanzibeica* (from the Western Sayan Mts.), European *W. geoides* and *W. trifolia*, North American *W. fragarioides* and *W. parviflora* was carried out. From two spacers of ITS region only ITS2 is recommended for molecular identification of plants by BOLD System. However, our data showed a high intraspecific and intra-individual variation of ITS2, together with a low number of systematic substitutions which do not allow distinguishing of closely related species. Thus, application of ITS2 region as a molecular marker for *Waldsteinia* spp. is very ambiguous. The detected levels of intra- and interspecific ITS1 and *trnH-psbA* variability allow using these molecular markers for identification of *Waldsteinia* species. The best results were observed in case of combining the ITS1 and *trnH-psbA* sequences together, which allowed to identify species in 100% cases according the 'best close match' test.

Keywords: DNA barcoding · *Waldsteinia* · Relict plant species · ITS1-ITS2 · *trnH-psbA*

© Springer Nature Switzerland AG 2019
I. Bychkov and V. Voronin (Eds.): *Information Technologies in the Research of Biodiversity*, SPEES, pp. 107–115, 2019.
https://doi.org/10.1007/978-3-030-11720-7_15

1 Introduction

At present, the methods applying the nucleotide sequences of short genome regions (DNA markers or molecular markers) for the biological species identification are actively developed. It becomes possible to fast and efficiently identify species, the determination of which by morphological criteria is difficult (Hebert et al. [6]. This principle formed the basis of the genetic database purposed the identification for all living organisms–Barcode of Life Data System (BOLD).

In our study, we estimated the possibility to identify some *Waldsteinia* Willd. species using DNA barcodes. *Waldsteinia* is a small genus of vascular plants belonging to the Rosaceae family and native to the temperate Northern Hemisphere. The genus includes around 5–9 species, however, systematic status of several representatives of the genus is still under discussion. In North America, the genus is represented by 3–4 species: *Waldsteinia fragarioides* (Michaux) Tratt. (=*W. fragarioides* subsp. *fragarioides* (Michaux) Tratt.), *W. parviflora* Small. or *W. doniana* Tratt. (=*W. fragarioides* var. *parviflora* (Small) Fernald, *W. fragarioides* subsp. *doniana* (Tratt.) Teppner), *W. lobata* (Baldw.) Torr. & A. Garay and *W. idahoensis* Piper. The question about status of *W. parviflora* is still open (Teppner et al. [10]; Weakley and Gandhi [13].

In the northern Eurasia, *Waldsteinia* is represented by *W. geoides* Willd. inhabiting the central and eastern Europe, and *W. ternata* (Steph.) Fritsch with highly fragmented distribution range. Originally *W. ternata* were described from the population on the Khamar-Daban ridge (the south of Baikal Siberia) but fragments of its range are also located in Middle Europe, south of Middle Siberia and in temperate East Asia. Later, the populations of *W. ternata* from the range fragments were accepted as separate species: *W. trifolia* Koch (=*W. ternata* subsp. *trifolia* (Koch) Teppner) for Europe, *W. tanzibeica* Stepanov for Western Sayan Mts, *W. maximowicziana* (Teppner) Probat. (=*W. ternata* ssp. *maximowicziana* Teppner) for East Asia. Nevertheless, the question of systematic rank of small species of *W. ternata* as well as North American *W. fragarioides* seems not available to be resolve only by morphology and geographical distribution (Teppner et al. [10].

The fact that some *Waldsteinia* species are considered as tertiary (nemoral) relicts and endangered species, for instance, South Siberian *W. ternata* and *W. tanzybeica*, and ornamental plants widely used in gardening (e.g. *W. geoides*, *W. ternata*) determines the importance of the study. The express method of species identification will be very useful for distribution monitoring of natural populations of endangered relict species as well as to control transportation and cultivation of ornamental clones. The difficulties in morphological identification of closely related *Waldsteinia* species require the development of approaches using DNA barcodes for the species identification.

For DNA barcoding of animal species the nucleotide sequence of the 5'-fragment encoding first subunit of cytochrome C oxidase (CO1 or *cox1*) is used as an universal marker (Hebert et al. [6]. Attempts to use universal markers for plants were less successful. The most promising markers for the plants DNA-barcoding are some fragments of the plastid genome (*rbcL, matK, trnH-psbA, atpF-atpH, rpoB, rpoC1, psbK-psbI*, etc.) and internal transcribed spacer region (ITS) of nuclear DNA (Hebert et al. [6]; Chase et al. [4]; CBOL Plant Working Group [3].

Internal transcribed spacer (ITS) is referred as the DNA region situated between the 18S and 26S genes of ribosomal RNA in transcribed region of polycistronic rRNA precursor. ITS consist of two spacer regions: ITS1 and ITS2 and 5.8S rRNA gene in between. For the phylogenetic studies the whole ITS region, including 5.8S gene is mostly used. However, for molecular identification of plants using BOLD system ITS2 region is recommended. In this study, we analyzed the possibility to use ITS1 and ITS2 regions separately together with *trnH-psbA* chloroplast intergenic spacer for DNA identification of *Waldsteinia* species.

2 Materials and Methods

2.1 Sampling

W. ternata populations were sampled from in the Khamar-Daban Ridge (floodplains of the Bezymyannaya, the Dulikha and the Snezhnaya rivers) and in the Eastern Sayan Mts. (floodplain of Belaya Zima river). *W. tanzybeica* is sampled in *locus classicus* in the Western Sayan Mts. (the Bolshoy Kebezh river floodplain). Samples of East Asian *W. maximowicziana* were collected in the Russian Far East (Nadezhdenskiy and Partizanskii areas, Primorsky Krai). European *W. geoides* was sampled in the Botanical Garden of Irkutsk State University (Irkutsk, Russia) where it had been cultivated since 2003. *W. trifolia* was sampled in the Botanical Garden of University of Wrocław, (Wrocław, Poland). Additional samples of *W. trifolia*, as well as North American *W. fragarioides* and *W. parviflora* were sampled from herbarium collection of the Komarov Botanical Institute RAR (LE; Saint-Petersburg, Russia). At least six individuals per local population were sampled randomly and for each specimen fresh leaves were collected and preserved in dry silica gel until DNA isolation. In case of sampling from herbaria, only one specimen from which herbarium sheet was collected.

2.2 Total DNA Isolation

Total gDNA was isolated from silica-dried leaf tissue following CTAB method (Doyle and Doyle [5] with modifications. Up to 100 mg of dried tissue from each specimen was grinded using automatic homogenizer MiniLys (Bertin technologies). 600 µl of CTAB extraction buffer (2% (w/v) cetyltrimethylammonium bromide, 100 mM Tris-HCl (pH 8.0), 20 mM EDTA (pH 8.0), 1.4 M NaCl, 3% (w/v) polyvinylpyrrolidone (PVP) (MW 40 kDa), 1% (v/v) β-mercaptoethanol) was added to the tissue powder, mixed well and incubated at 60 °C for 1 h, 500 µL of chloroform-isoamyl alcohol mixture in the ratio of 24:1 (v/v) were added to the homogenate, and the samples were centrifuged at 14,000 g for 15 min at 4 °C. The upper aqueous phase was transferred to a new tube, the equal volume of chloroform-isoamyl alcohol mixture was added and centrifuged as before. DNA was precipitated by addition of 0.8 volumes of isopropanol to 1 volume of water phase followed by incubation for at least 1 h at −20 °C. DNA was pelleted by centrifugation at 14,000 g for 15 min at 4 °C, washed twice by 70% ethanol, air-dried at room temperature, and then resuspended in nuclease-free water preheated at 60 °C.

2.3   PCR and Sanger Sequencing

The complete ITS region was amplified using the forward ITS1-P2 (Utelli et al. [11] and the reverse ITS4 [14] primers. For *trnH-psbA* region amplification, combination of forward and reverse primers (Sang et al. [8]; Tate and Simpson [9] which successfully applied for *W. fragarioides* (Burgess et al. [2] were used. PCR was performed using both GoTaq Flexi DNA Polymerase kit supplied with green buffer (Promega) and Q5 High-Fidelity 2X Master Mix (New England BioLabs) according to the manufactory protocol. PCR products were visualized in 1.5% agarose gel stained by ethidium bromide after electrophoresis, gel-purified using the GeneJET Gel Extraction Kit (Thermo Fisher Scientific), ligated into pTZ57R/T (Thermo Fisher Scientific) or pMiniT 2.0 (New England BioLabs) plasmid vectors followed to cloning in TOP10 *E. coli* competent cells or directly sequenced. Bacterial cloning was used for the detection of low copy number variants of ITS region. Isolated plasmids and original amplicons were sequenced by Sanger method using BigDye Terminator Cycle Sequencing kit v. 3.1 (Applied Biosystems) on 3500 Genetic Analyzer (Applied Biosystems).

2.4   Sequences Alignment and Data Analysis

The original forward and reverse sequences were edited using SnapGene Viewer software v.2.6.2. Additionally, to the originally obtained nucleotide sequences the sequences deposited in GenBank were also used in the analyses. For *W. fragarioides* the sequences with accession numbers AH006923.2 (ITS) and HQ596900.1 (*trnH-psbA*) and for *W. geoides* the sequence with accession number AJ302362.1 (ITS) were used for the alignment. The multiple alignment of nucleotide sequences by MUSCLE algorithm were carried out in MEGA software v. 7.0.16. Barcodes of nucleotide sequences were generated using on-line tool – DNA Barcode Generator (Bio-Rad). Sequences of ITS1, ITS2 and *trnH-psbA* regions were analyzed both separate and combining together. The efficiency to use each molecular marker for species barcoding estimated by the presence/absence of 'barcoding gaps' between the interspecific and intraspecific distances which were evaluated using frequency histograms based on the uncorrected paired p-distances and by the 'best close match' test calculated in Tax-onDNA software v. 1.8 (Meier et al. [7]; Bolson et al. [1]. The long insertion/deletions regions in alignments were considered as one evolutionary event for p-distance calculation between nucleotide sequences. The 'best close match' test was used to determine the closest correspondence of a target sequence with all of the others in a set of aligned data, there by establishing a similarity limit based on the distribution of the frequencies of the intraspecific and interspecific distances (Bolson et al. [1]. The efficiency of each marker in species identification was estimated by portion of sequences which were classified as "correct", when the genetic distances between the target sequence and other sequences of the same species fit within 95% of the calculated limit.

3   Results and Discussion

The multiple alignments of 45 ITS nucleotide sequences of seven *Waldsteinia* species showed that the length of ITS1 region is 229 bp and ITS2 contain 207 bp, no gaps were found in the both DNA regions. The multiple alignment of ITS1 sequences contigs of *Waldsteinia* species visualized as barcodes is shown on a Fig. 1. Each contig was obtained by multiple alignment of intraindividual nucleotide sequences of ITS1 region.

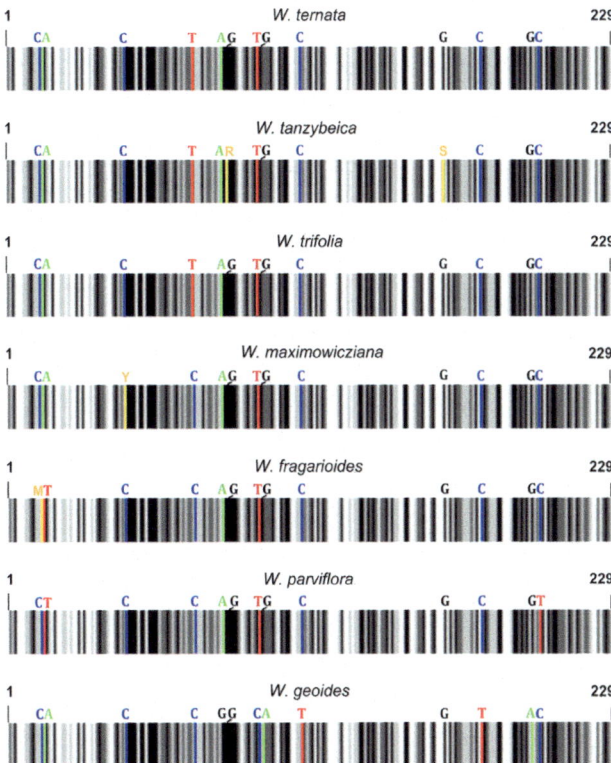

Fig. 1. DNA barcodes of several *Waldsteinia* species based on multiple alignment of ITS1 nucleotide sequences contigs. Four type of nucleotides are presented as dashes colored by different shades of gray range (A – white, C – light gray, T – dark gray, G – black). Colour dashes and letters indicates the polymorphic sites with interspecies differences. Degenerate nucleotide characters are marked by yellow (M is A *or* C; R is A *or* G; S is G *or* C; Y is C *or* T). Numbers indicate the fragment length in base pairs.

The number of polymorphic sites in ITS1 region was 13 (5.7%) including 4 degenerative characters in three of seven species and all of them were indicated the interspecies differences. Nine from those thirteen polymorphic sites contained only single-valued substitutions, which can unambiguously discriminate 4 species and one species aggregate combining *W. ternata*, *W. tanzybeica* and *W. trifolia*. Using degenerative characters in *W. tanzybeica* barcodes generated on the base of ITS1 sequences and indicating the intra-individual variability allow to single out this species from the species aggregate.

In ITS2 region 12 polymorphic sites were found, however, only two of them contained single-valued substitutions, which allow to distinguish only two species, i.e. *W. geoides* and *W. trifolia* (the data is not presented). The other 10 sites are contained degenerative nucleotides indicating intraspecific variability, the most part of which fitted on intra-individual polymorphism, detected by molecular cloning. The high numbers of the degenerate sites based on intraspecific polymorphism together with a low level of single-valued interspecies substitutions do not allow to identify species unambiguously. Moreover, five species have identical variants of ITS2 region. The presence of a barcoding gap for individual or combined regions represented by the frequency distribution of the genetic distance between intraspecific sequences and between interspecific sequences was observed only for ITS1 region but not for ITS2 region (Fig. 2).

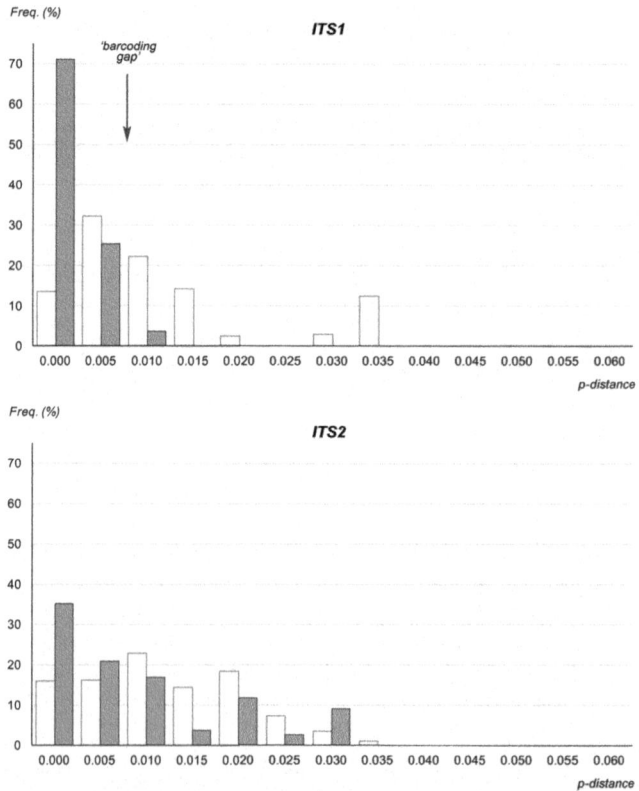

Fig. 2. The frequencies of pairwise intraspecific (grey bars) and interspecific (white bars) divergences based on uncorrected paired p-distances for ITS1 and ITS2 DNA regions.

ITS1 presented the barcode gap with 86.5% of pairwise interspecific p-distances greater than 0 and 54.2% – greater than 0.005 and 71% of pairwise intraspecific p-distances equal 0 and 96.4% – lower than 0.005.

The results of the 'best close match' test showed correct identification of nearly 56% for ITS1 and of 43% for ITS2. Thus, despite that for molecular identification of plants using BOLD system ITS2 region is recommended, our data showed that using of ITS2 region as a molecular marker for *Waldsteinia* spp. is very ambiguous. The higher efficiency of ITS1 in comparison with the ITS2 region is shown also in a number of other species (Wang et al. [12].

The length of multiple alignment of 27 *trnH-psbA* nucleotide sequences was 375 bp including long gaps (111 bp in total). Twelve polymorphic sites with single-valued substitutions and 5 sites with informative gaps were observed. The *trnH-psbA* region presented the barcode gap with 88% of pairwise interspecific p-distances greater than 0.005 and 79% of pairwise intraspecific p-distances lower than 0.005 (Fig. 3).

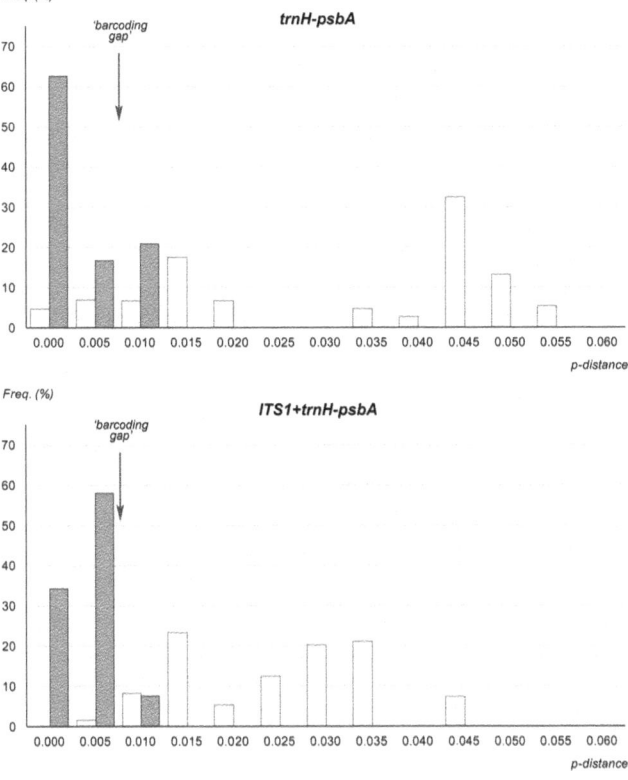

Fig. 3. The frequencies of pairwise intraspecific (grey bars) and interspecific (white bars) divergences based on uncorrected paired p-distance for *trnH-psbA* DNA region individually and combined with ITS1.

The results of the 'best close match' test for *trnH-psbA* region showed correct identification of nearly 59%. Although *trnH-psbA* provided the highest values of interspecific genetic distances, with a range varying from zero to a maximum of 0.055, this was combined with a high frequency (nearly 5%) of p-distance equaled 0 at the interspecific level.

Nonetheless, complete discontinuity was not found between the intra- and inter-specific ranges of p-distances for *trnH-psbA* as well as for ITS markers. On the next step of the study we analyzed the markers showed the highest efficiency (ITS1 and *trnH-psbA*) together. The best barcode gap for combined region instead of individual markers tested was found with 98.3% of pairwise interspecific p-distances greater than 0.005 and 92.3% of pairwise intraspecific p-distances lower than 0.005 (Fig. 3). The results of the 'best close match' test for ITS1+*trnH-psbA* showed correct identification in 100% cases.

4   Conclusion

Despite that ITS2 region is recommended for molecular identification of plants using BOLD system, its applicability for *Waldsteinia* species is very ambiguous. A high intraspecific and intra-individual variation of ITS2 region, together with a low number of systematic substitutions do not allow to distinguish closely related *Waldsteinia* species. ITS1 and *trnH-psbA* molecular markers are more promising for this purpose. The best results were obtained when ITS1 and *trnH-psbA* markers analyzed as a combined data set. It allowed to identify species in 100% cases according the 'best close match' test.

Acknowledgments. The study of the Far Eastern, the European and the North American populations was supported by the Russian Science Foundation (project No. 17-74-10074) and the study of Siberian populations was supported by Russian Foundation for Basic Research (projects No 16-34-60135_mol_a_dk and 16-05-00783). The research was done using the equipment of The Core Facilities Center "Bioanalitika" and collections of The Core Facilities Center "Bioresource Center" at Siberian Institute of Plant Physiology and Biochemistry SB RAS (Irkutsk, Russia). The authors also thank IRKU, IRK and LE herbaria and Botanical Garden of Irkutsk State University allowing to collect samples, Prof. N.V. Stepanov, Prof. N.S. Probatova and Dr. E.A. Pimenova for their help with sampling.

References

1. Bolson, M., de Camargo Smidt, E., Brotto, M.L., Silva-Pereira, V.: ITS and trnH-psbA as efficient DNA barcodes to identify threatened commercial woody Angiosperms from Southern Brazilian Atlantic rainforests. PLoS ONE **10**(12), e0143049 (2015)
2. Burgess, K.S., Fazekas, A.J., Kesanakurti, P.R., Graham, S.W., Husband, B.C., Newmaster, S.G., Percy, D.M., Hajibabaei, M., Barrett, S.C.H.: Discriminating plant species in a local temperate flora using the rbcL+matK DNA barcode. Methods Ecol. Evol. **2**, 333–340 (2011)
3. CBOL Plant Working Group: A DNA barcode for land plants. PNAS **106**(31), 12794–12797 (2009)

4. Chase, M.W., Salamin, N., Wilkinson, M., Dunwell, J.M., Kesanakurthi, R.P., Haidar, N., Savolainen, V.: Land plants and DNA barcodes: short-term and long-term goals. Philos. Trans. R. Soc. B Biol. Sci. **360**, 1889–1895 (2005)

5. Doyle, J.J., Doyle, J.L.: A rapid DNA isolation procedure for small quantities of fresh leaf tissue. Phytochem. Bull. **19**, 11–15 (1987)

6. Hebert, P.D.N., Cywinska, A., Ball, Sh.L, deWaard, J.R.: Biological identifications through DNA barcodes. Proc. R. Soc. B Biol. Sci. **270**, 313–321 (2003)

7. Meier, R., Shiyang, K., Vaidya, G., Ng, P.K.L.: DNA Barcoding and taxonomy in Diptera: a tale of high intra- specific variability and low identification success. Syst. Biol. **55**, 715–728 (2006)

8. Sang, T., Crawford, D.J., Stuessy, T.F.: Chloroplast DNA phylogeny, reticulate evolution, and biogeography of *Paeonia* (*Paeoniaceae*). Am. J. Bot. **84**, 1120–1136 (1997)

9. Tate, J.A., Simpson, B.B.: Paraphyly of *Tarasa* (*Malvaceae*) and diverse origins of the polyploid species. Syst. Bot. **28**, 723–737 (2003)

10. Teppner, H., Schuehly, W., Weakley, A.S.: The chromosome numbers of Waldsteinia (Rosaceae- Colurieae) in North America. Phyton **48**(2), 225–238 (2009)

11. Utelli, A., Roy, B., Baltisberger, M.: Molecular and morphological analyses of European *Aconitum* species (*Ranunculaceae*). Plant Syst. Evol. **224**, 195–212 (2000)

12. Wang, X.-C., Liu, Ch., Huang, L., Bengtsson-Palme, J., Chen, H., Zhang, J., Cai, D., Li, J-Q.: ITS1: a DNA barcode better than ITS2 in eukaryotes? Mol. Ecol. Resour. **15**, 573–586 (2015)

13. Weakley, A.S., Gandhi, K.N.: Recognition of three taxa of eastern North American "*Waldsteinia*" and their appropriate names when incorporated into *Geum* (*Colurieae: Rosaceae*). J. Bot. Res. Inst. Tex. **2**(1), 415–418 (2008)

14. White, T.J., Bruns, T,. Lee, S., Taylor, J.: Amplification and direct sequencing of fungal ribosomal RNA genes for phylogenetics. In: Innis, M.A., Gelfand, D.H., Sninsky, J.J., White, T.J. (eds.) PCR protocols: a guide to methods and applications, pp. 315–322. Academic Press, San Diego (1990)

Technology of Information and Analytical Support for Interdisciplinary Environmental Studies in the Baikal Region

Igor V. Bychkov[1], Gennady M. Ruzhnikov[2], Roman K. Fedorov[1],
Yurii V. Avramenko[1], Alexander S. Shumilov[1], Alexei O. Shigarov[1],
Alla V. Verhozina[3], Natalia V. Emelyanova[4],
and Andrei A. Sorokovoi[4(✉)]

[1] Matrosov Institute for System Dynamics and Control Theory SB RAS, Irkutsk,
Russia
[2] Irkutsk Scientific Center SB RAS, Irkutsk, Russia
[3] Siberian Institute of Plant Physiology and Biochemistry SB RAS, Irkutsk,
Russia
[4] V.B, Sochava Institute of Geography SB RAS, Irkutsk, Russia
fedorov@icc.ru

Abstract. The technology of creating a geoportal information-analytical environment for supporting scientific research of nature management in the Baikal region is considered on the basis of a service-oriented paradigm and the integrated use of thematic and spatial data accumulated by the scientific community.

Keywords: Information-analytical environment · Services · Geoportal · Spatial data · Nature management

1 Introduction

Nature management is an object of interdisciplinary scientific research, since it represents the totality of all forms of interaction between society and nature, combining not only the processes of human use of natural conditions and resources, anthropogenic impact on nature and its modification by this influence, but also the reproduction of the natural environment. With rational nature management, the most complete satisfaction of needs for material wealth is achieved while maintaining the ecological balance and the possibilities for restoring the natural resource potential.

In the field of nature management and environmental protection, the emphasis is placed on:

- solving problems of the development of the economic state complex, which takes into account the environmental and natural-geographical conditions of specific areas;
- the achievement on each particular territory of the quality of the habitat, which corresponds to the system of assessments of the genetic health of the population;

I. Bychkov and V. Voronin (Eds.): *Information Technologies*
in the Research of Biodiversity, SPEES, pp. 116–124, 2019.
https://doi.org/10.1007/978-3-030-11720-7_16

- restoration and preservation of biosphere balance, genetic fund of wildlife;
- rational use of the entire natural resource potential.

An important component of modern nature management is its monitoring, which makes it possible to develop a system for managing environmental management processes. Integration, processing of spatio-temporal monitoring data and their comprehensive analysis help to formulate new solutions in this area.

Scientific studies of the state and dynamics of the natural ecosystems of the Baikal region and their components that have been actively developing in recent years are based on databases of data, knowledge and services localized in the institutes of the SB RAS, universities, territorial authorities and government. Open information exchange of support for nature management in the region is not established, which makes it difficult to conduct complex interdisciplinary research and implementation of their results in the management of territorial development.

Modern development of information technologies, increasing the data transfer rate, development and standardization of the browser interface and services, the creation of data centers (data centers), the development of the methodology of "cloud computing" allow us to transfer information and analysis systems to the Internet and implement openness, scalability, provision of common classifiers, the availability of data and services for their processing, etc. [1, 2].

The development of an open information and analytical environment (IAS) for support of interdisciplinary environmental studies in the Baikal region on the basis of modern information technologies is relevant for working out integrated use of thematic and spatial data accumulated by the scientific community on environmental management issues.

IAS, as the environment for integration, storage, distributed processing of WPS services and integrated data use should provide:

- creation of thematic and spatial data using crowdsourcing methodology (crowdsourcing, crowd and sourcing), similar to the Wikipedia projects, NASA Clickworkers, eBird, Peer-to-Patent, etc., but unlike them, IAS data should be created in a relational, normalized form, oriented to automatic processing;
- creation of a prototype of the ETL Web service for integration of unstructured tabular data;
- data storage in data centers, which will ensure reliability, online access from any device connected to the Internet;
- regulated work with data in order to preserve intellectual property;
- data exchange and their integration into common resources;
- viewing IAS data and distributed processing on the basis of WPS-services and their combinations [3].

The development of the IAS allows creating a single information space supporting complex interdisciplinary studies of natural ecosystems for obtaining new knowledge and their implementation in managing the territorial development of the region.

The main object for the approbation of environmental components is the study of the transformation of the flora of the territories with a change in the level of urbanization and the development of technology to assess the spatial and temporal changes in biodiversity from various factors. The distributed (cloud) computing environment provides both the input and storage of user data on the biodiversity of the Baikal region, and the analysis of these data using WPS services.

In order to partially fill the IAS, studies were conducted in the following areas:

- estimation of integration trends in the development of the transport system of the Baikal region using spatial analysis services [4];
- forecasting, using thematic services, changes in the floristic composition of the Baikal region, depending on the socio-economic development of the region [5];
- development of methods and systems for visualizing maps to show the dynamics of changes in the state of the Baikal region [2];
- development of thematic maps of the main water objects of the territory and points of monitoring of quantitative and qualitative characteristics of water resources;
- creation of a series of socio-economic maps for urban development and urbanization [6]:
- spatial-temporal dynamics of the regional system of cities;
- differentiation of cities by levels and types of socio-demographic development.

2   Information-Analytical Environment for Monitoring of Nature Use Objects

Service-oriented paradigm, OGC standards allow performing at the new level interdisciplinary scientific researches of nature management of the Baikal region. The main component of IAS is a typical geoportal (Fig. 1), which, as an access point, supports the exchange of spatial information between users (researchers), and services (Fig. 2) [1].

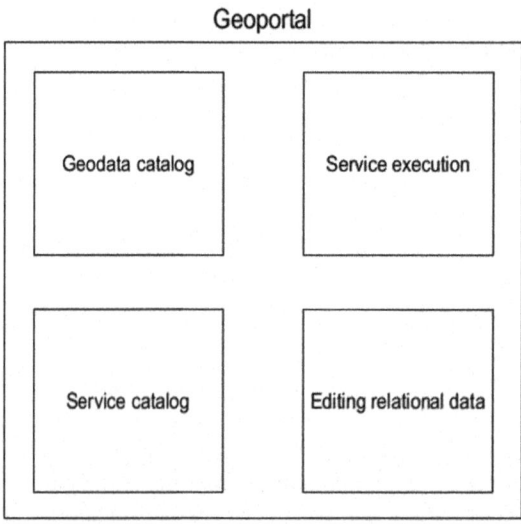

Fig. 1. The structure of a typical geoportal.

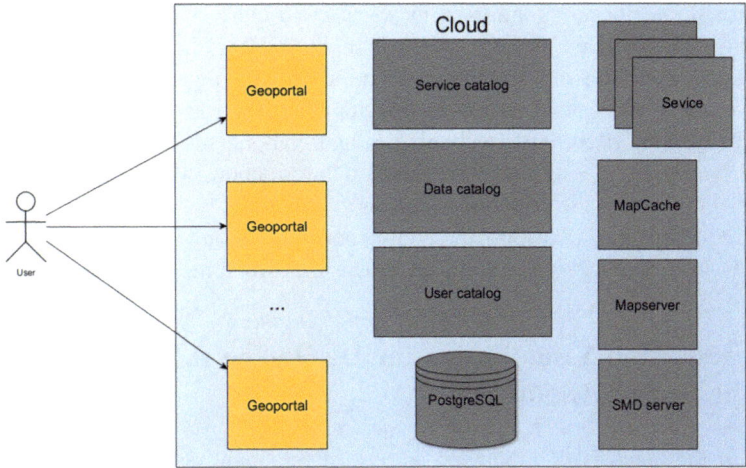

Fig. 2. The architecture of the cloud environment for monitoring.

The main components of a typical geoportal:

1. Catalogs of data and services, subsystems for starting services and displaying data.
2. System for input and editing of relational data with spatial attributes, providing access to data services within the WPS standard.

Interdisciplinarity of environmental management tasks argues for the need to create a cluster of thematic geoportals that support the exchange of spatial and thematic data and Web services for their processing based on OGC standards, which increases network load balancing and simplifies the work of users.

For rapid deployment of a new instance of the geoportal a template has been created. To store spatial data, the PostgreSQL DBMS is used. The use of a common DBMS for all cluster geoportals makes it possible to simplify data exchange and use various user tables in queries. Mapserver and SMD Server are required to display spatial data in accordance with the WMS standard. Mapserver displays the constantly changing data in the PostgreSQL database. SMD Server displays static data at high speed without using caching systems.

To automate the search for services and data, a structural specification is developed in JSON format, which defines the input data of services, describes the structure of relational tables. This specification can be converted into the required standards (XSD, WFS, WSDL-S). In addition to field descriptions (name, data type), the structural specifications contain:

- metadata in accordance with ISO 19115;
- units for numeric data;
- possible values for enumerated data types;
- widgets.

The widgets implement a user interface for adding, editing, and displaying field data.

Structural specifications are used by the user to create tables that query the corresponding data services and are stored in a specialized directory and arranged in hierarchies. A mechanism has been developed for converting relational data to DBF or SHAPE (if there are spatial attributes) for applying them as input to WPS services. Unlike the existing ones, these technologies and components use structural specifications for the representation and processing of information, which allows them to be applied flexibly to a wide range of problems.

The IAS implements forming, integrating and transmitting data, launching services, organizing the computing process in the cloud infrastructure, presenting results, etc.

3 Tools of Map Visualization for Displaying the Dynamics of the Baikal Region State

As part of the services of input and editing of relational data, the style editor was developed to specify the mapping of geoportal layers based on the WMS standard. The style editor uses the metadata catalogs for spatial data and cartographic symbols (name, display, keywords and description).

The style editor for each layer allows you to specify several display classes based on attribute values (Fig. 3). For each class, you can define your own mapping using symbols from the cartographic symbols directory. The styles are translated to the Mapserver display setting file [1, 2].

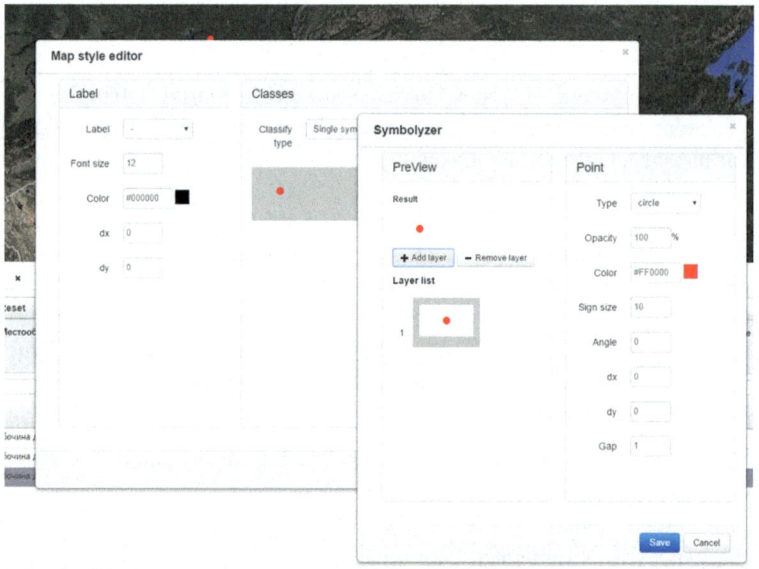

Fig. 3. The style editor.

Within the geoportal, it is possible to include data presentation styles in a set of service metadata, which allows them to be used automatically, both for displaying service parameters and for the results of services.

4 Web Service ETL for Integration of Unstructured Tabular Data

Web-service for integration of unstructured tabular data is based on rules of analysis and interpretation of arbitrary labeled tables containing data on the nature use of the Baikal region that are collected and accumulated in unstructured form in spreadsheets with an arbitrary structure [1, 2].

The prototype of the RESTful Web service ETL-integration of unstructured table data (RWS-SSDC, RESTful Web Service for SpreadSheet Data Canonicalization) is implemented. Web-service provides: loading of initial table data (arbitrary spreadsheets, CRL-programs, YAML-specifications of categories) to the server; setting up and launching a data transformation system from arbitrary Excel spreadsheets based on the execution of CRL-programs; generation of output JSON data (original arbitrary table, recovered semantic relations, output canonical table); saving the output data on the server and providing access to it (Fig. 4.). The Web service is implemented as a servlet for the Java EE execution platform.

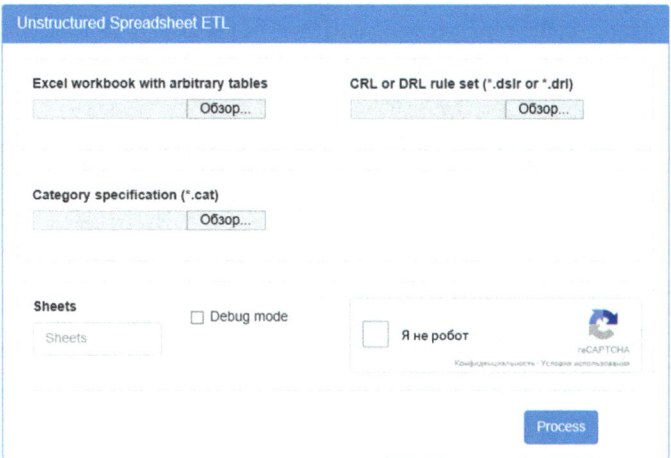

Fig. 4. User web interface.

In the data transformation system of spreadsheets, the analysis and loading of the source data is performed using the free Apache POI system. As a result, objects are created representing the table, its cells, rows and columns. They are added as facts into the working memory of the rule execution system. Optionally, one or more categories described with the YAML language can be loaded into the system, also forming relevant facts in the working memory of Drools. In this case, the parsing of YAML descriptions is performed in a free SnakeYAML system. Data structures that represent cells, occurrences, labels, and categories are implemented as Java classes in accordance with the JavaBeans specification conventions.

Each fact is an instance of one of these classes. Rules for analyzing and interpreting tables are expressed in a special CRL language. As a rule inference system, Drools software is used. The facts loaded in the working memory are compared to the compiled rules. As a result, new facts are created that describe the semantics of the table: occurrences, labels, categories and relations between them. The canonical table is generated from them.

Web-client access to RWS-SSDC provides service management, visualization and interactive work with the output JSON data received from it. The Web client is implemented as a JavaScript application using the Bootstrap platform.

The developed service was used to export demographic data from statistical reports in a PDF format to a geoportal.

5   Service of Spatial Analysis and Evaluation of Formation Trends of the Transport System of the Baikal Region

To estimate the impact of the transport system on nature in the Baikal region, a WPS-service for building transport access maps was created. The investigated territory is divided by a regular grid. In each cell, the time of movement to this point from the nearest municipal formation is calculated. It is assumed that in the beginning traffic is carried out by road on the road network, and then (if necessary) on foot. The following

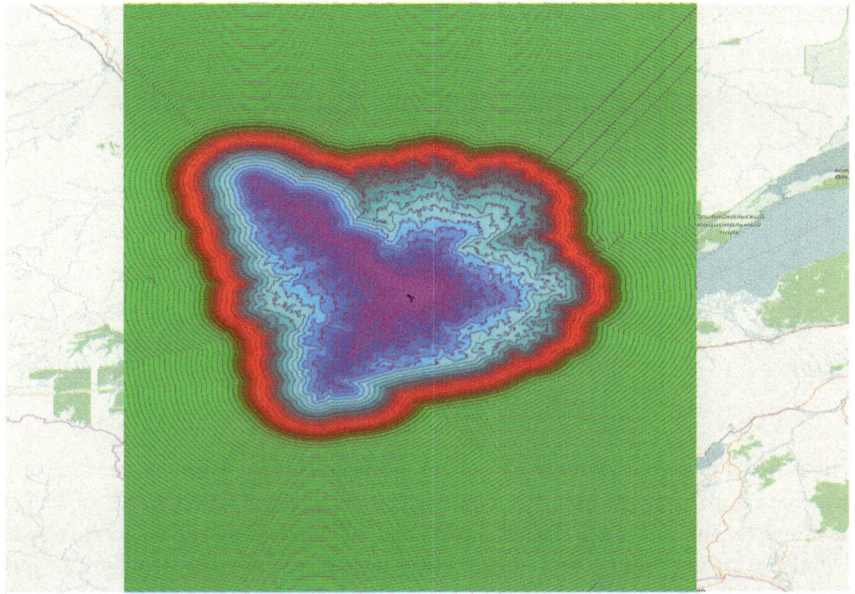

Fig. 5. Isolines of transport accessibility in 30 min increments.

data are transferred to the service entrance: the boundaries of municipal formations in the form of polygons, a network of roads in the form of a set of polylines, the extent of the map being created, and the size of the cell. At the output of the service we get a regular grid in GeoTIFF format (Fig. 5).

6 Conclusion

The information and analytical environment for support of interdisciplinary studies of nature management in the Baikal region have been created. Its advantages are the openness, distribution, wide application of the methods for processing thematic and spatial data on the nature use of the Baikal region in the form of Web services. The databases of thematic spatial data are partially filled and maps of the main water objects and points of monitoring of water resources characteristics, as well as socio-economic development of cities and urbanization of municipalities:

1. Map "Hydrological posts on the territory of the Russian part of the Lake Baikal basin".
2. Map "Differentiation of cities in the Baikal region by types of social-demographic development".
3. The map "The permanent population of the Baikal region".
4. Map "The degree of urbanization of municipalities of the Baikal region".
5. Map "Indicators of environmental burden and environmental costs in the subjects of the Baikal region".

Acknowledgements. This work was supported by: RFBR (grants №: 18-07-00758-a, 17-57-44006-монг_a, 17-47-380007-р-а, 16-07-00411-a, 16-07-00554-a); Programs of the Presidium of the Russian Academy of Sciences No. 27; Integration Programs of SB RAS, INC SB RAS). The results were obtained using the Integrated Information and Computing Network of the Irkutsk Scientific and Educational Complex.

References

1. Bychkov, I., Ruzhnikov, G., Khmelnov, A.: Integration of information and analytical resources and processing of spatial data in problems of managing territorial development. Novosibirsk: Izd. Siberian Branch of the Russian Academy of Sciences. 369 p. (2012)
2. Bychkov, I., Ruzhnikov, G., Khmelnov, A., Fedorov, R., Paramonov, V., Shigarov, A., Fereferov, E., Gachenko, A., Mikhailov, A., Shumilov, A., Avramenko, Yu., Madzhara, T.: Infrastructure of information resources and technologies for creating information-analytical systems of territorial management. Izdat. SB RAS, 242 p. (2016)
3. Fedorov, R., Shumilov, A.: The graph of dependencies for service compositions using JavaScript scenarios. Geography and natural resources. № 6 - Spec. Release, pp. 160–163 (2016)
4. Dugarova, G., Bogdanov, V.: Analysis of the current state of roadside service in the Irkutsk region. ECO. №2. 117–126 (2015)

5. Verkhozina, A., Fedorov, R., Kazanovsky, S., Shumilov, A., Krivenko, D., Murashko, V.: Informational and analytical system on phytorrhages of the Baikal Siberia. Izvestiya Irkutsk State University. Series "Biology. Ecology". vol. 9, No. 3, 9–16 (2016)
6. Vorobiev, N., Emelyanova, N., Vorobiev, A., Valeeva, O.: Settlement, population dynamics and human potential of the Baikal region. Geography and natural resources. Spec. Release, pp. 168–178 (2016)
7. Kuzmin, S., Nevzorova, I., Cherkashin, E., Shamanova, S.: Geoinformation mapping based on the model of relief plastics and the possibility of its use in geomorphological analysis. Geoinformatics **2**, 19–34 (2016)

Geoinformational Web-System for the Analysis of the Expansion of the Baikal Crustaceans of the Yenisei River

A. V. Andrianova[1,2(✉)] 🆔 and O. E. Yakubailik[1,3] 🆔

[1] Institute of Computational Modelling SB RAS, Akademgorodok 50/44,
660036 Krasnoyarsk, Russia
{andrav,oleg}@icm.krasn.ru
[2] Research Institute of Ecology of Fishery Reservoirs, ul. Parizhskoi Kommuny
33, 660097 Krasnoyarsk, Russia
[3] Siberian Federal University, Svobodny pr. 79, 660041 Krasnoyarsk, Russia

Abstract. The technologies and related software are developed for considered problem. The geospatial database is generated and filled with the results of own long-term hydrobiological field studies, it has become an integral part of the geoportal of ICM SB RAS, which was formed by a separate thematic section. The focus is on the results of extensive field studies of the Yenisei implemented in 2015 and 2016. As to database content, the information about the quantitative distribution of zoobenthos (animals inhabiting the ponds bottom), in particular endemic Baikal amphipods, in the area from Yenisei river headwaters to its delta was used. It is revealed that endemic Baikal amphipods has spread far beyond not only down, but also up on the Yenisei river. After the commissioning of the Krasnoyarsk hydroelectric power station, their share in the total zoobenthos biomass is increased by 10 times. *Gmelinoides fasciatus* crustacean is especially active; it massively has populated the area of the Upper Yenisei river below the Sayano-Shushensky reservoir. The density and the fraction of crustaceans in the zoobenthos in the area of the Angara–Podkamennaya Tunguska has increased over the past 15 years.

Keywords: Spatial distribution · Geographic information system · Zoobenthos · Amphipods · Geoportal · Web mapping · Yenisei

1 Introduction

Yenisei River is the main Siberian river, it is one of the seven largest rivers in the world and it is the most high-water river in Russia. It is known that the emergence of large hydraulic structures causes a prolonged, and sometimes irreversible, destabilization of aquatic ecosystems. The construction of Krasnoyarsk hydroelectric power station (HPS) in the 70s of the last century caused global changes in hydrological, hydrochemical and hydrobiological regimes in the Yenisei River. The Yenisei River in the downstream of the HPS does not freeze in winter over 100–300 km from the dam; the influence of hydropower station on the ice regime of the river is traced to the mouth of the river Podkamennaya Tunguska. Yenisei hydropower engineering has caused global changes in

© Springer Nature Switzerland AG 2019
I. Bychkov and V. Voronin (Eds.): *Information Technologies in the Research of Biodiversity*, SPEES, pp. 125–130, 2019.
https://doi.org/10.1007/978-3-030-11720-7_17

hydrobiological communities. There was a change in the dominant forms of phyto-plankton, an enrichment of its species composition and an increase in the total number of algae due to runoff from the upper reach; the amount of phytobenthos and phytoperi-phyton, which became even a hindrance in the work of water intakes, has sharply increased [1]. The change in the hydrological regime had a significant impact on popu-lations of sturgeon and other valuable fish species, significantly violating their ranges [2].

Because of the HPS construction zoobenthos (invertebrate animals living in water bodies on the ground and in its depth) in the Yenisei had global changes, especially in the HPS downstream. The stoneflies and blackflies have almost disappeared from the benthic fauna, the density and number of caddisflies and mayflies species have sig-nificantly decreased. Quantitative characteristics of zoobenthos in the area from the dam to the Angara's estuary have greatly increased: quantity–by more than 2 times, biomass–by 5 times. The growth of indicators is determined, firstly, by the spread of Gammaridae from Lake Baikal through the Angara river upstream of the Yenisei, while their proportion in the total zoobenthos biomass increased by 10 times [3, 4].

In this context, the spatial distribution of amphipods throughout the Yenisei River and the analysis of the dynamics of invasive processes are of some interest. Nowadays, the problem of invasions of alien species belongs to one of the important directions of fundamental and applied research. However, piece of information of this kind of research in our country is still insufficiently developed. In Russia, there is a lack of Internet resources dedicated to invasive species [5].

The ecosystem of the Yenisei has accumulated over a long period a vast array of diverse biotic and abiotic data that are valuable for analysis. The use of new methods and technologies of data processing, such as geoinformation and cartographic mod-eling, provides the opportunity to obtain additional information on spatial features of the hydrobionts distribution, helps in the search of relationships with various envi-ronmental factors.

2   Materials and Methods of Hydrobiological Research

The presented materials are the continuation of studies of the condition of the Yenisei fodder resources started by us in the early 2000s [3, 4, 6]. At this stage, modern data (2015 and 2016) on the state of the bottom fauna in general, and in particular on the distribution of the Baikal amphipods in the Yenisei from the mouth to the delta inclusive, have been obtained. Samples were taken at a depth of 1 m; only in the Yenisei delta it was possible to remove the soil from depths of up to 14 m. When collecting hydrobiological material, the depth, water temperature, dissolved oxygen content, transparency, flow rate, degree of macrophyte overgrowth, and soil type were determined. For this research 178 quantitative samples of bottom fauna were collected in the Yenisei and analyzed. Collection and analysis of the material was carried out by conventional hydrobiological methods.

In the Upper Yenisei, there were 15 stations for collection of hydrobiological material: 2 of them are located within the Republic of Tuva; 3–lower than the Sayano-Shushensky reservoir (from the city Sayanogorsk to the city Minusinsk); and 10–on the section from the Krasnoyarsk HPS dam to the mouth of the river Angara. In the Middle

Yenisei the area from the mouth of the river Angara to the village Surgutiha was explored–21 stations, in the Lower Yenisei zoobenthos was collected at 12 stations– from the city Dudinka to the delta, including the Brekhov Islands.

3   Results of the Hydrobiological Research

In the Yenisei we discovered several species of amphipods, but throughout the whole river there was only *Gmelinoides fasciatus* Stebb. In the Upper Yenisei it was the single and rare representative of the higher crustaceans. Below the Krasnoyarsk dam, the species composition of the amphipods increased due to *Philolimnogammarus viridis* Dybowsky, *Gammarus sp., Ph. cyaneus* Dybowsky, *Pallasea cancelloides* Gerstfeldt, *Eulimnogammarus verrucosus* Gerstfeldt. In the Middle Yenisei after the Angara outlet, this complex of amphipods persisted, but *Gammarus sp.* Was changed by *Micruropus* sp. In the Lower Yenisei the greatest value was acquired by *Pontoporeia affinis* Lindstrom.

The quantitative advantage in the investigated areas of the Yenisei was taken by *G. fasciatus* and, to a lesser extent, *Ph. viridis*, in the delta they were replaced by *P. affinis*. The number of amphipods in different zones of the Yenisei varied significantly (Table 1). At the highest investigated site (the Republic of Tuva) under conditions of high flow velocity and large rocky-pebble soil, the amphipods are presented extremely poor: only *G. fasciatus* was found individually. The maximum density of the amphipods was below the Sayano-Shushensky reservoir near cities Sayanogorsk and Minusinsk (3.8 thousand ind./m^2 and 10.4 g/m^2), while their share in the zoobenthos averaged 70% of the population and 53% of the biomass (Table 1). Further downstream, the number of amphipods decreased and in the Lower Yenisei fell to an average

Table 1. Number (in numerator, ind./m^2) and biomass (in denominator, g/m^2) of zoobenthos and amphipods in the river Yenisei.

The site of the Yenisei	Zoobenthos	Amphipods	G. fasciatus	Ph. viridis
Upper Yenisei:				
the Republic of Tuva	612 ± 93	Unique	Unique	Absent
	4.20 ± 1.68			
Sayanogorsk–Minusinsk	5501 ± 2186	3809 ± 1682	3398	411
	19.5 ± 4.09	10.4 ± 3.90	6.47	3.96
Krasnoyarsk HPS–the mouth of Angara	2769 ± 509	962 ± 311	705	177
	10.5 ± 2.51	4.97 ± 1.72	1.29	1.90
Middle Yenisei:				
The mouth of Angara–the village Surgutiha	1423 ± 150	643 ± 121	486	60
	6.44 ± 0.73	3.22 ± 0.53	1.72	0.86
Lower Yenisei:				
Dudinka–delta	2234 ± 419	288 ± 81	67	201*
	8.55 ± 1.66	0.60 ± 0.13	0.22	0.32*

Примечание: * - number and biomass of *P. affinis*

of 0.3 thousand specimens/m^2 with a biomass of 0.6 g/m^2 (13% abundance and 7% biomass of benthic fauna). A sharp decrease in the density of the amphipods on the lower reaches of the Middle Yenisei can be explained by the hydrological features of this section, due to the fact that after the confluence of the Podkamennaya Tunguska river the channel slope is reduced by half, that leads to a decrease in flow velocity and to the accumulation of silt sediments [7].

After 15 years, from the moment of our previous large-scale studies of the zoobenthos of the Yenisei, it was found out that the quantitative characteristics did not change significantly on the reach Krasnoyarsk HPS–Angara. However, the density of both species statistically increased significantly below the mouth of the Angara, especially of *G. fasciatus*, whose size became 4 times larger and the corresponding biomass became 7 times larger. The proportion of crustaceans in zoobenthos has also increased.

4   Geoinformation Database

Research and development carried out in the field of geoinformation support of hydrobiological monitoring tasks is a continuation of the work begun in 2015 [8]. On the basis of our own data of expedition studies, we created a geospatial database with observational results (Fig. 1), which is placed on the geoportal of the ICM SB RAS in a separate thematic section (http://gis.krasn.ru/go/n5p8).

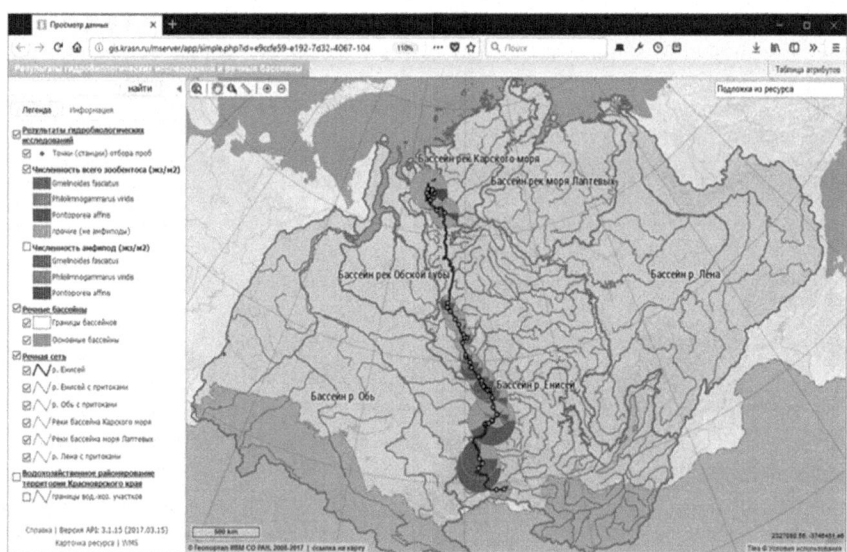

Fig. 1. Web interface of the ICM SB RAS geoportal: a river network of the Yenisei basin, selection points for hydrobiological samples, research results.

Geoportal provides tools for visualization and processing of geodata, access to them from outside applications based on cartographic web services [9]. The initial filling of the geo-information system was composed with the materials of expeditionary research conducted in the early 2000s on the section from the dam of Krasnoyarsk HPS to the mouth of the river Podkamennaya Tunguska [8, 10]. At present, the information resource is supplemented with up-to-date information on the quantitative distribution of zoobenthos, including amphipods, all over the Yenisei.

The developed geoinformation database with the results of hydrobiological monitoring is available through the standard software web services of the geoportal. Thanks to these interfaces, users are given the opportunity of analytical processing and presentation of hydrobiological monitoring data, export to programs such as Microsoft Excel for further analysis.

Based on the prepared geospatial data, geo-referencing of the observation points of hydrobiological monitoring to kilometer markers along the Yenisei fairway was made., i.e. the "coordinate system of the river" is formed. The use of such a coordinate system makes it possible to evaluate various parameters along the course of a river. Also, a new mechanism was proposed and implemented for displaying contextual information about selected objects on the map–based on templates. Templates allow you to flexibly configure the design and content of an information pop-up window with data about the selected map object: to change the style design–color, font settings, etc., to place photos, place interactive elements on the form–menus, selectors, etc., to run external scripts for requests to external data [11].

Within the considered research stage, considerable attention was paid to the actualization, systematization and structuring of information on the hydrography of the Yenisei basin. Feasibility and significance of this work is determined by the prospects of complex hydrobiological and hydrographic modeling, the use of spatial analysis methods in geoinformation systems.

With the help of GIS software, a set of multiscale digital cartographic data on the river network in a GIS-format was prepared. The developed hydrographic data is not only layers with water objects (rivers, lakes, etc.), but also geospatial information of a special type - databases and GIS layers that allow to make different analysis and modeling, visualization of information. The initial data included information from the State Water Cadastre (water register), various tables and handbooks on the characteristics of water bodies from publicly available open resources, and Russian and foreign multi-scale digital cartographic data on the river network and water-collecting territories from various sources.

At further stages of the work, it is planned to involve methods of geoinformation modeling. Geo-positioned hydrobiological information can be juxtaposed with other elements of the natural environment - plant cover, soil types, state environmental monitoring data on pollution of the territory, and so on. The river basin in this context becomes a system-forming factor. The study of the various characteristics of the ecosystem based on the basin principle links the diverse manifestations of biological life in the territory into one.

5 Conclusion

Creation of the geospatial database with the results of field researches and implementation of geoinformation web-system for information and analytical support of the hydrobiological monitoring greatly expands opportunities in the analysis and representation of geodata, and form the basis for interdisciplinary studies. For the subject area under consideration, this approach is particularly relevant, taking into consideration the significant spatial distribution of information.

References

1. Ponomareva, Y.A., Ivanova, E.A.: Ratio between living and dead cells and the size structure of the Yenisei River Phytoplankton downstream of the Krasnoyarsk hydroelectric power station. Contemp. Probl. Ecol. 9(5), 706–717 (2016)
2. Zadelenov, V.A.: Modern state of sturgeon fish (Acipenseridae) populations and their nutrition base in the Yenisei basin. Contemp. Probl. Ecol. 7(3), 287–291 (2000). (In Russ.)
3. Gladyshev, M.I., Moskvicheva, A.V.: Baikal invaders have become dominant in the Upper Yenisei benthofauna. Dokl. Biol. Sci. 383(1–6), 138–140 (2002)
4. Andrianova, A.V.: Dynamics of Yenisei zoobenthos evolution in the downstream of Krasnoyarsk hydroelectric power station. Tomsk State Univ. J. Biol. 1(21), 74–88 (2013). https://doi.org/10.17223/19988591/21/6. (In Russ.)
5. Dgebuadze, YuYu., Petrosyan, V.G., Bessonov, S.A., Dergunova, N.N., Izhevskiy, S.S., Maslyakov, VYu., Morozova, O.V., Tsarevskaya, N.G.: General concept of development of a problem-oriented internet portal on alien species invasion on the Russian Federation territory. Russ. J. Biol. Invasions 2, 9–21 (2008)
6. Andrianova, A.V., Yakubailik, O.E., Shulepina, S.P.: The usage of GIS technologies for the analysis of spatio-temporal dynamics of the Baikal amphipods in the Yenisei river. In: Proceedings of the XI Congress of Hydrobiological Socety at RAS, Siberian Federal University, Krasnoyarsk, pp. 17–18 (2014). (In Russ.)
7. Greze, V.N.: Food resources for fish in the Yenisei river and their utilization, vol. 41, 236 p. Izvestiya Vniohr, Moscow (1957) (In Russ.)
8. Andrianova, A.V., Yakubailik, O.E.: Geographic information web system providing the hydrobiological monitoring on the example of Yenisei river zoobenthos. Comput. Technol. 21(1), 5–14 (2016). (In Russ.)
9. Yakubailik, O., Kadochnikov, A., Tokarev, A.: Applied software tools and services for rapid Web GIS development. Inform. Geoinformatics Remote Sens. (SGEM 2015). 1(2), 487–496 (2015). https://doi.org/10.5593/SGEM2015/B21/S8.060
10. Andrianova, A., Shaparev, N., Yakubailik, O.: Geoinformation support and web technologies for problems of hydrobiological monitoring of Yenisei river. MATEC Web Conf. 79, 01056 (2016). https://doi.org/10.1051/matecconf/20167901056
11. Kadochnikov, A.A., Yakubailik, O.E.: Service-oriented web applications for spatial data processing. Novosibirsk State Univ. J. Inform. Technol. 13(1), 37–45 (2015). (In Russ.)

Some Problems of Regional Reference Plots System for Ground Support of Remote Sensing Materials Processing

Alina Bavrina[1,2], Anna Denisova[1], Lyudmila Kavelenova[1(✉)],
Eugeny Korchikov[1], Oksana Kuzovenko[1], Nataly Prokhorova[1],
Darya Terentyeva[1], and Victor Fedoseev[1,2(✉)]

[1] Samara National Research University, Samara, Russia
lkavelenova@mail.ru, vicanfed@gmail.com
[2] Image Processing Systems Institute, Branch of the Federal Scientific Research
Centre "Crystallography and Photonics" of Russian Academy of Sciences,
Molodogvardeiskaya st. 151, Samara 443001, Russia

Abstract. The landscape complexes state assessment, including natural and anthropogenically transformed ecosystems, is the basis for the organization of rational nature management and biological diversity conservation. A significant amount of remote sensing data already accumulated and constantly replenished can become the basis for a large-scale spatial and temporal analysis of the local ecosystems. However, the necessary condition for that, in our opinion, is the use of regionally verified ground survey data along with remote sensing data. Such ground survey data may be represented by regional reference plots of ecosystems having different states (natural, disturbed, converted, and regenerated). As the characteristics associated with species composition and vegetation condition become more detailed, the importance of ground-based measurement increases. On the example of the forest-steppe and steppe areas of Middle Povolzhye (Samara region) as an ecotone with a mosaic combination of preserved natural ecosystems, agroecosystems, urban areas, spontaneously recovering ecosystems, we can demonstrate the importance of regional reference plots network for ground support of remote sensing data processing. In this paper, we discuss some methods of combining regional reference plots with remote sensing data during the ecosystem analysis. These methods are the result of the joint investigations of two research groups of ecologists and remote sensing data processing specialists.

Keywords: Remote sensing · Reference plots system · Natural and anthropogenically transformed ecosystems · Data clustering · Sentinel-2

1 Introduction

Remote sensing of the Earth from space (RS) is one of the most successful and dynamically developing innovative industries. The areas of academic and applied use of remote sensing data include meteorological and topo-geodetic support, and also a wide range of problems in environmental management, agriculture, target cadasters maintaining, hydrometeorology, climate studies, emergency monitoring situations of

© Springer Nature Switzerland AG 2019
I. Bychkov and V. Voronin (Eds.): *Information Technologies
in the Research of Biodiversity*, SPEES, pp. 131–143, 2019.
https://doi.org/10.1007/978-3-030-11720-7_18

natural and technogenic nature, ecological monitoring, cartography, a fundamental study of the Earth and its evolution. Unfortunately, the priority of our country in these fields has been largely lost in recent years.

In recent years the development of the concept of "ecological services" [1–5], which addresses the ways to optimize the use of natural resources at the local, regional and global levels, urgently requires serious attention to information support of activities with the mandatory use of remote sensing data [1]. Many indicators of the ecosystem state show a fairly close relationship with remote sensing data obtained by measuring reflected signals. In particular, this refers to the determination of the primary productivity of ecosystems by evaluating the fluorescence of chlorophyll [1]. Some other indicators such as climate regulation are indirectly related to the remote sensing data (specifically, to the remote estimations of surface temperature). Finally, a number of important indicators (food production, raw materials, the state of the natural components of ecosystems) cannot be adequately reflected when using only remote sensing data. This case requires the joint use of remote sensing data and other information sources, including field observations [1, 6].

In the last decade, there has been a sharp increase in the number of works linking remote sensing data with the ecosystem components state assessment and the provision of ecological services [7–11]. A systematic review [1] by Barbosa et al. revealed 5920 publications since 1960 indexed in Scopus and Web of Science, associated with the term "ecological service". From this list, 211 papers consider the use of remote sensing data. The confluence of these works with the countries demonstrated the absolute leadership of the USA and China, while no papers from Russia were found. However, this fact does not mean the absence of such works, their number is quite large. For example, we indicate a few papers [12–15], which can make it possible to comprehend the breadth of the circle of specialists and the research problems in this field in Russia.

When assessing the state of terrestrial and aquatic ecosystems based on remote sensing data, including those related to the activities of producers, the calculation of various vegetation indices is often used. This is done mainly to separate the green vegetating vegetation from other types of underlying surface (primarily from the soil cover and water surface), and also to assess its current state. In order to ensure that the numerous tasks of natural resources use optimization with the help of remote sensing are adequately resolved, it is necessary to continue and expand the work in the development of software and methodological support for obtaining more accurate and reliable ecological state estimates.

In addition to the actual results of space imagery, it is necessary to obtain interfaced ground information blocks during calibration. The aerospace polygon is a territory representative of the physical and geographical conditions for a region with a typical set of natural complexes and their modification as a result of both natural and anthropogenic impacts of varying degrees. Within the territory of the test site, test (reference) areas are marked out, on which stationary methodical studies of the most typical objects and conditions are carried out [15].

The list of the main research tasks on the instrumentation includes, in addition to solving the methodological and technical bases of remote sensing, also collecting information on the characteristics of objects and phenomena, determining their significant deciphering and indicator features, accumulating statistical data and creating a bank of deciphering features, improving the methods for processing remote sensing data. It also carries out the development of experimental scientific and pro-industrial work and training to disseminate experience. We emphasize specially that the main directions of using polygons are the development of etalons of topographic and thematic interpretation, the creation of a catalog of deciphering features, and subsatellite synchronous measurements [15].

Considering the possibilities of combining the data of the ground-based survey of the state of biological diversity in the natural and anthropogenically transformed ecosystems of the Samara region, at the first stage of the work we encountered the lack of regional test and measurement ranges, the quantitative characteristics of which could be adequately used for working with RS materials. This, as well as the unification of various specialists within the framework of the Samara National Research University, was the starting point for the beginning of our joint work.

2 Problems and Approaches in Combining Remote Sensing Data and Ground Observations for Samara Region

2.1 General Assumptions and Approaches

The problem, which limits the possibilities of direct extrapolation of the already available experience of domestic and foreign works in ecosystems state monitoring in the Samara Region using RS data, is connected with the fragmented nature of the distribution of different ecosystems types here, also as their mosaic interpenetration and relatively limited area. This is the result of the historical development of natural landscapes, and in recent decades - the strengthening of the anthropogenic press. In particular, for a number of districts in the Samara region, the share of agrocenoses exceeds 90% (Fig. 1), and fragments of natural ecosystems remain "in the residual form". That is why there was a need to work in a limited space, with the selection of many standards, with the construction of original or adapted to regional features of recognition systems.

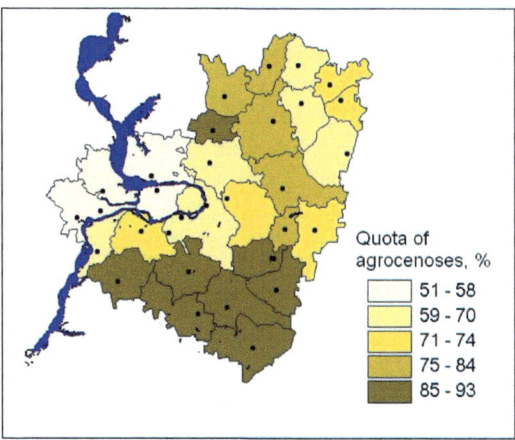

Fig. 1. Modern state of Samara region districts according to agrocenoses quota in their total area.

At the first stage, after studying foreign and domestic experience, we have formulated an algorithm for selecting sites for ground-based polygons and their surveys by the specifics of their ecosystem affiliation. It consists in selecting, by a detailed analysis of available cartographic and other material, of model plots for conducting a comprehensive field survey. The information set collected on the polygon have to include the data on landscape confinement features, localization in various relief forms, as well as information on the of vegetation and soil cover state estimated by scales, percentages or other quantitative units including specially developed forms. A set of indicators is formed taking into account the ecosystem belonging of the polygon sites to the steppe, forest or another type. The information on the ground survey of polygons, after the desk processing and generalization, is then used together with RS data.

To select a priority RS data source for ecosystem state assessment, we took into account the following factors. Typical sizes of the analyzed objects (areas of steppe or forest with a homogeneous vegetation composition) range from several tens to several hundred meters across, so high- and medium-resolution space images are acceptable to detect and classify such objects. One more requirement for remote sensing data is the presence of the near-infrared channel because this band is the most valuable in vegetation analysis. In addition, given the transience of the growing season, an important feature to satellite mission is short revisit time. As a result, the choice was made between the data of Resurs-P satellites (Geotone sensor, available data for 2016), Sentinel-2A and 2B (MSI sensor), Landsat-8 (OLI sensor). Their comparative characteristics are listed in Table 1.

Table 1. Comparative characteristics of remote sensing data sources.

Satellite and sensor	Resolution	Swath width	Revisit time	Access
Resurs-P Geotone	3–4 m (5 channels)	38 km	3 days	Limited
Sentinel-2 MSI	10 m (4 channels), 20 m (6 channels)	290 km	5 days or less	Free
Landsat-8 OLI	30 m (9 channels)	185 km	16 days	Free

This table shows the advantage of Sentinel-2 over Landsat-8 in all major indicators. As for the Geotone data, despite the theoretical shooting periodicity in 3 days, we have access to the only image archive for 2016. In this archive, for most areas, only 1–2 cloudless images during the growing season are available, excepting city districts shot up to four times. As a result, Sentinel-2 satellites (A and B) are the most appropriate data source.

2.2   Data Selection for a Particular Territory

As an example, we analyzed the available Sentinel-2 imagery for April and May 2018 for the area near Bolshaya Chernigovka village. It includes two protected sites of steppe vegetation "Urochishche Mulin Dol" and "Fescue-feather virgin steppe area", and also some lands in their neighborhood. This area is characterized by a high degree of transformation of natural ecosystems into agrocenoses, with the preservation of fragments of steppes, meadows, forests. Taking into account the volume of available images that meet the goals of our work, and the confinement of valuable natural sites, in the summer of 2018, reference areas should be selected, grouped into steppe, bush-steppe, forest communities. From the point of view of biological diversity, rare species of plants, including those with different protection status, are confined to them (Fig. 2). Shrub communities are also refugia for animals that choose shelter here among the visible steppe space and agrocenoses.

Fig. 2. The total view and some rare plant species of some regional ecosystem: upper row– steppe community with *Stipa* sp. predominance; *Ornitogalum fischeranum* Krasch; low row– shrubs community in steppe; *Gladiolus tenuis* M.Bieb. (both plants included in Red Book of Samara Region).

For the selected area, we found 11 Sentinel-2 images for April–May 2018, including 7 cloudless, which cover various phases of plant growth. The analysis of these images has shown that the combined use of remote sensing data covering different phases of plant growth allows the separating vegetation types better. This feature is partly demonstrated in Fig. 3. Taking into account these assumptions, we decided to use for data clustering and classification multispectral composite images made from Sentinel-2 data of 10-meter resolution for different dates. In addition to the four

Sentinel-2 channels (blue, green, red and infrared), we added the NDVI channels for each date. Thus, for the considered territory, when using data for April–May 2018, such a composite image of 35 channels can be used.

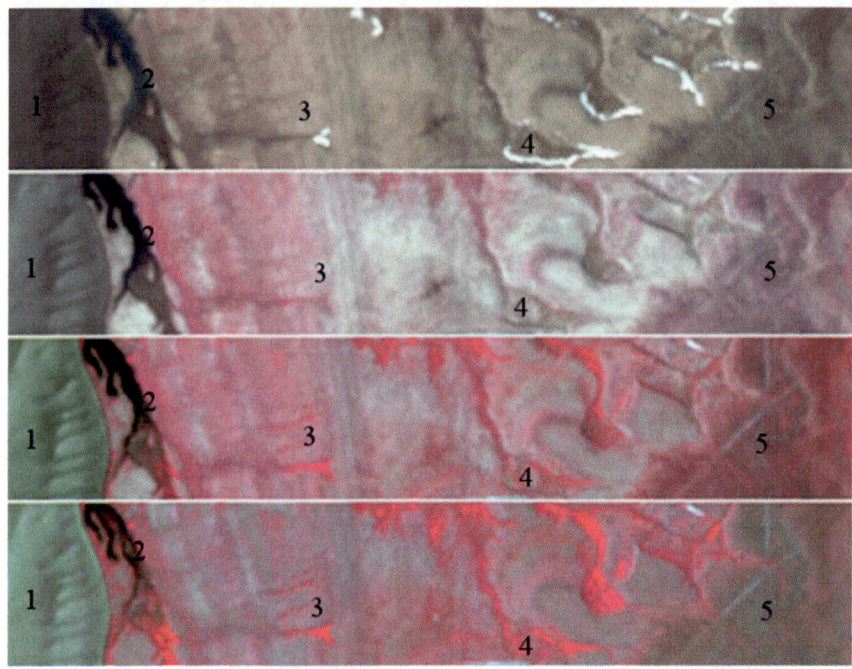

Fig. 3. Steppe vegetation in the Sentinel-2 images dated 17 April 2018, 2 May 2018, 14 May 2018 and 24 May 2017 (top – down). The fragments contain: 1–field, 2–water, 3–forb-feather grass steppe, 4–bushes, 5–fescue-wormwood steppe. The visualized channels are NIR, Red, and Green.

3   Remote Sensing Data Clustering for Ground-Truth Data Preparation

An important step in the natural ecosystem analysis is the selection of reference sites (ground-truth). During this step, we need to follow some considerations. Firstly, the reference areas should be sufficiently homogeneous (in terms of species diversity, topography and soil cover). Secondly, they should have a certain minimum area. And finally, the plant communities in the selected sites should be phenologically different in the studied time period and potentially detectable by existing methods of image processing.

Therefore, in this study, we considered a controlled approach for reference site selection. It consists in merging different remote sensing images into composites and clustering them. It may help to determine the potentially separable areas of vegetation within the study area. This method makes it possible to construct a preliminary map of homogeneous areas of vegetation. Their boundaries and species may be specified later during ground surveys.

Therefore, we performed the following steps of the preliminary RS image analysis:

– constructing a data composite for different dates;
– its cluster analysis;
– analysis of clustering results to select potential reference sites;

For the composite construction Sentinel-2 data, acquired at the period of May 1–30 2018, were used. The overall amount of cloudless images was equal to 5. With the use of these images, the composite of 5 channels of NDVI indices at different dates (May 2, 14, 22, 24 and 27) was produced.

The resulting composite, thus, characterizes the presence and density of green vegetation at each image point. The account of different-time observations in the considered period allows to analyze the differences in the occurrence of phenological phases of the first shoots and the formation of a thick growing cover. Examples of visualization of the constructed composite are shown in Fig. 4.

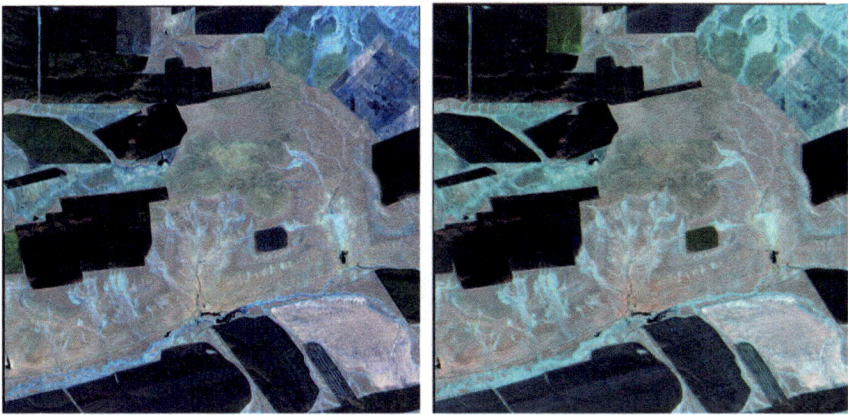

Fig. 4. Visualization of 5-channel composite NDVI image (size 743 × 772) used for clustering (two images represent different channel triples).

Clustering was performed by the general EM clustering algorithm with undefined number of clusters [16]. The algorithm evaluates the number of clusters using a hierarchical histogram of the input image. The advantages of this clustering algorithm are the adaptive cluster number determination and the effective multivariate data processing. The algorithm defines cluster centers as the modes of the global data distribution. Thus, it produces the set of the most probable cluster parameters according to the generalized Gaussian mixture model.

Each cluster corresponds to one of the major vegetation types that are presented in the scene. The image is labeled during the clustering process by the numbers of clusters. Therefore, each connected region with the same cluster number can be interpreted as an area with the same dominant vegetation type. The distinction between vegetation types is based on the fact that the set of NDVI values indicate the degree of green coverage development during the observation period. In other words, the vegetation area with the same vegetation type will have the similar NDVI values during this period. As only the images captured in May 2018 were used, the distinguishable major vegetation types differ from each other by the onset time of such phenological phases as shooting and tillering.

General EM clustering algorithm has one parameter. It is a minimum frequency value (MFV) for the histogram cell. This parameter stands for the minimum probability of the histogram cells which are regarded as nonempty cells. The cells with lower frequency value are removed from the histogram representation. The MFV value regulates the generalization properties of the clustering process and the accuracy of the global distribution assessment. The higher value of the histogram frequency cell corresponds to the higher value of generalization and, at the same time, to the lower value of the cluster parameters accuracy. To define the appropriate MFV value we conducted an experiment with different MFV values equal to 280, 560 and 1120 points per cell, that correspond to the probabilities of 0,0005, 0,001 and 0,002. After visual analysis of the obtained clustering results, the MFV value equal to 280 was selected for further ground-based polygons determination. The results of the clustering using different MFV values are shown in Fig. 5.

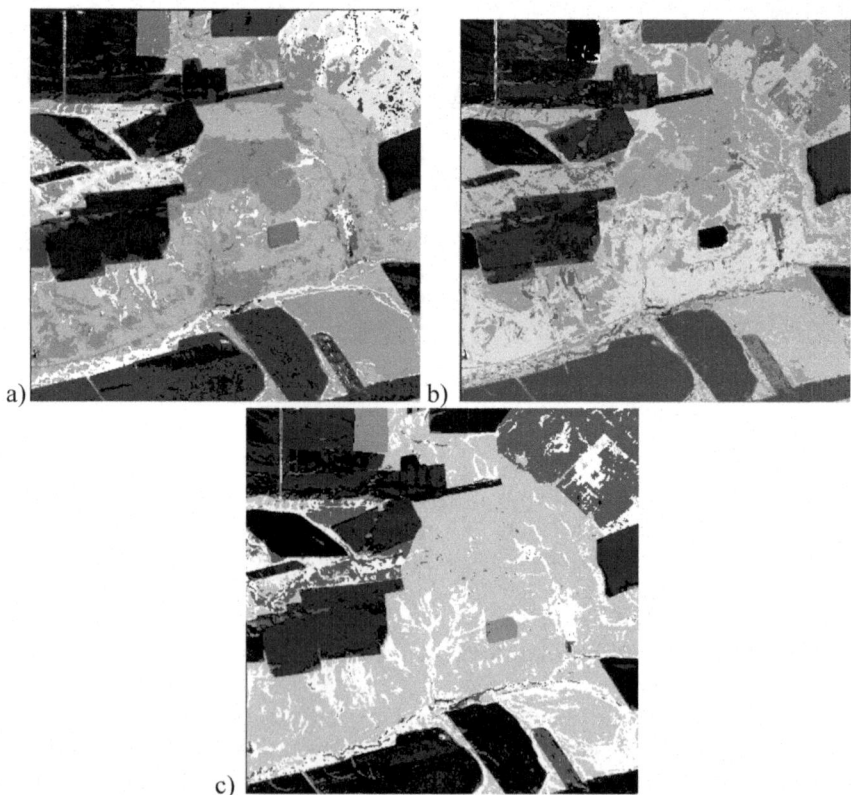

Fig. 5. Clustering results for different MFV values: (a) 280, (b) 560, (c) 1120.

Clustering with MFV values equal to 280, 560 and 1120 results in 12, 9 and 5 clusters correspondingly. Visual analysis of the original image and clustering borders showed that the best clustering parameter value is 280, because it results in more precise clustering of natural vegetation and the greater MFV values lead to less number of clusters with more general composition i.e. the greater values of MFV make unable to distinguish different natural vegetation classes from each other. Otherwise, the fewer values of MFV lead to excessive clustering with very small connected regions that are very difficult to verify and use in the classification process.

The obtained clustering results for MFV = 280 were converted into a vector form using GDAL and ArcGIS software. A fragment of the territory with the vectorized boundaries is shown in Fig. 6. The primary analysis of the obtained clustering data together with the remote sensing images showed that the clusters 0–3 correspond to the areas of agricultural fields, i.e. do not belong to the natural vegetation. Therefore, these agricultural clusters as well as regions of small area were removed from the dataset to decrease the amount of data to be analyzed. The rest regions with their cluster labels were converted into *.kml format and were given to the ecologists for the further analysis.

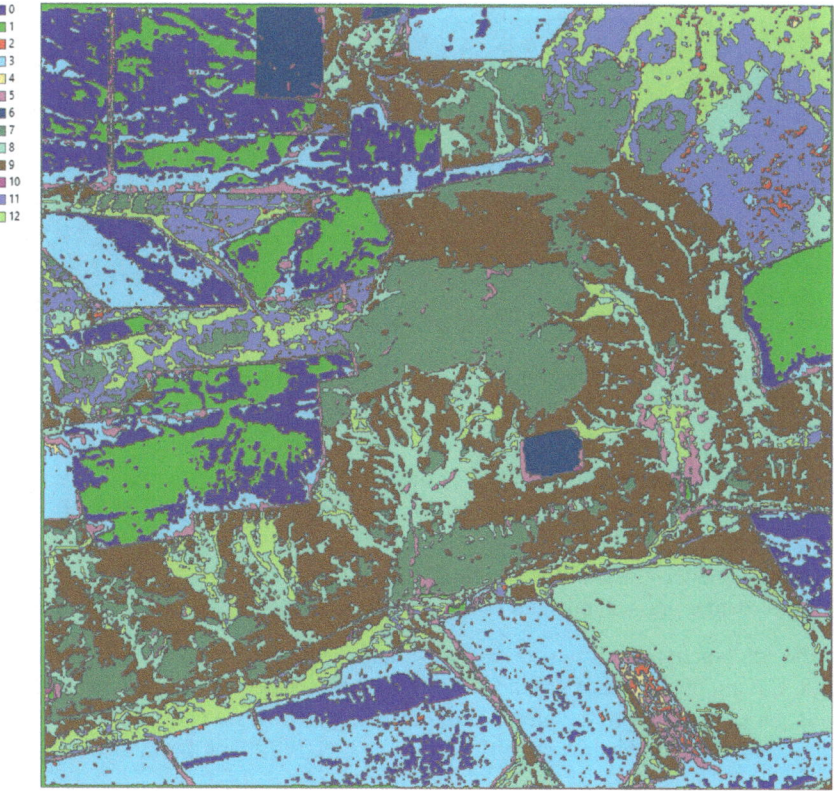

Fig. 6. Vectorized clustering results for MFV = 280.

The 12 reference areas (Fig. 7) were defined within the largest areas corresponding to different clusters of natural vegetation. Ecologists examined these areas. As a result, the following main classes of natural vegetation were identified:

Fig. 7. Fragment of the territory in the vicinity of the studied reference areas (Sentinel-2, 27 May 2018).

(1) fescue-wormwood steppe (point 1, cluster 7),
(2) meadow steppes in depressions and bushes along rivers (points 4, 5, 8–10, clusters 8 and 12);
(3) fescue-feather grass steppe (point 7, cluster 9),
(4) forb-feather grass steppe and saline meadows (points 2, 3, 11, and 12, cluster 11).

The combination of clustering results with a RS image for 27 May 2018 is shown in Fig. 8. Having point data of ground surveys without specifying the exact boundaries of vegetation areas and with the presence of clustering results, we constructed a sample of ground-truth data for the area of 2500 hectares, which is 72% of the studied image area.

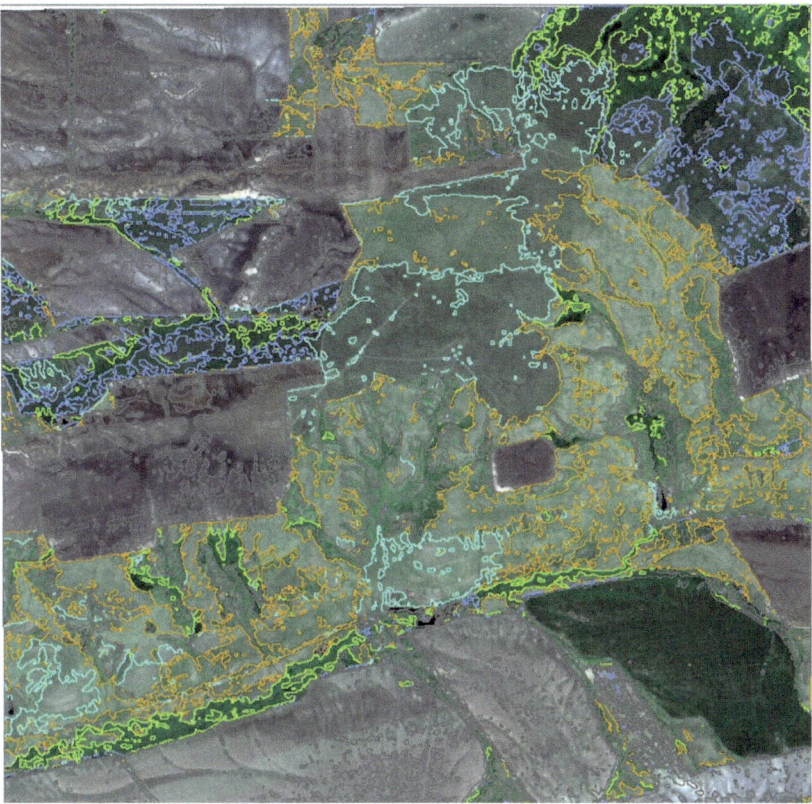

Fig. 8. The result of combining the clustering of the dominant classes of natural vegetation and the RS image (Sentinel-2, 27 May 2018).

4   Conclusion

The article deals with the problems of building and developing a system of ground-truth data for the verification of remote sensing data for the forest-steppe and the steppe areas of Middle Volga region. The peculiarity of this ecotone, which is a mosaic combination of natural and anthropogenic ecosystems, does not allow to use pure ground observations data accumulated for other territories of Russia.

To optimize the work of ecologists on the ground, as well as to better coordinate the location of the reference areas with the Sentinel-2 images selected as the main data source, we developed an approach based on the preliminary clustering of the composite NDVI data for different dates within the growing season. This allows us to distinguish different plant communities by their chronological differences in the growth phases.

After the clustering, we selected some cluster centers as reference points, and as a result of their ground survey we selected the main classes of plant communities which are specific for the considered territory. Thus, the proposed approach makes it possible to effectively use the knowledge and capabilities of specialists in different fields.

Further research will be conducted in the following directions:

- exclude from consideration (for clustering and classification) the areas relevant to human activities,
- use Sentinel-2 data for the whole vegetation period (from April to October),
- train a spatial classifier on the allocated reference sites,
- investigate the obtained classifier in another territory.

Acknowledgments. This work was supported by subsidy (08-08), allocated in accordance with the agreement of 02/26/2018, No. 074-02-2018-294, provided for state support of Samara National Research University in order to improve competitiveness accounting among the world's leading research and educational centers.

References

1. de Araujo Barbosa, C.C., Atkinson, P.M., Dearing, J.A.: Remote sensing of ecosystem services: a systematic review. Ecol. Ind. **52**, 430–443 (2015)
2. Costanza, R., Kubiszewski, I.: The authorship structure of "ecosystem services" as a transdisciplinary field of scholarship. Ecosyst. Serv. **1**, 16–25 (2012)
3. de Groot, R.S., Alkemade, R., Braat, L., Hein, L., Willemen, L.: Challenges in integrating the concept of ecosystem services and values in landscape planning, management and decision making. Ecol. Complex. **7**, 260–272 (2010)
4. de Groot, R.S., Wilson, M.A., Boumans, R.M.J.: A typology for the classification, description and valuation of ecosystem functions, goods and services. Ecol. Econ. **41**, 393–408 (2002)
5. Boyd, J., Banzhaf, S.: What are ecosystem services? the need for standardized environmental accounting units. Ecol. Econ. **63**, 616–626 (2007)
6. Chambers, J.Q., Asner, G.P., Morton, D.C., Anderson, L.O., Saatchi, S.S., Espírito-Santo, F. D.B., Palace, M., Souza, C.: Regional ecosystem structure and function: ecological insights from remote sensing of tropical forests. Trends Ecol. Evol. **22**, 414–423 (2007)
7. Murray, N., Phinn, S., Clemens, R., Roelfsema, C., Fuller, R., Murray, N.J., Phinn, S.R., Clemens, R.S., Roelfsema, C.M., Fuller, R.A.: Continental scale mapping of tidal flats across East Asia using the landsat archive. Remote Sens. **4**, 3417–3426 (2012)
8. Atzberger, C., Rembold, F., Atzberger, C., Rembold, F.: Mapping the spatial distribution of winter crops at sub-pixel level using AVHRR NDVI time series and neural nets. Remote Sens. **5**, 1335–1354 (2013)
9. Mozumder, C., Reddy, K.V., Pratap, D.: Air pollution modeling from remotely sensed data using regression techniques. J. Indian Soc. Remote Sens. **41**, 269–277 (2013)
10. Prabakaran, C., Singh, C.P., Panigrahy, S., Parihar, J.S.: Retrieval of forest phenological parameters from remote sensing-based NDVI time-series data. Curr. Sci. **105**, 795–802 (2013)
11. Nagendra, H., Lucas, R., Honrado, J.P., Jongman, R.H.G., Tarantino, C., Adamo, M., Mairota, P.: Remote sensing for conservation monitoring: assessing protected areas, habitat extent, habitat condition, species diversity, and threats. Ecol. Ind. **33**, 45–59 (2013)
12. Labutina, I., Baldina, E.: Use of remote sensing data for monitoring of protected area ecosystems. Tutorial. WWF Russia, Moscow (2011). (In Russian)
13. Classifier of thematic tasks of assessment of natural resources and the environment, solved using the Earth remote sensing materials. Baikal Center LLC, Irkutsk (2008). (In Russian)

14. Sukhikh, V., Zhirin, V.: Application of scanner space images in the inventory of reserve forests. In: Remote methods in forest management and forest inventory: instruments and technologies, pp. 92–97. Sukachev Forest Institute SB RAS, Krasnoyarsk (2005)
15. Gruzinov, V.: Methodology for assessing the "topographic information" of remote sensing data on test areas. Geod. Cartogr. **6**, 33–35 (2006). (In Russian)
16. Denisova, A., Sergeyev, V.: Using hierarchical histogram representation for the EM clustering algorithm enhancement. In: Proceedings of the 10th International Symposium on Image and Signal Processing and Analysis, pp. 41–46 (2017)

Late Cenozoic Lagomorphs Diversity in Eurasia

M. A. Erbajeva[1,2(✉)] (iD)

[1] Geological Institute, Siberian Branch RAS, Ulan-Ude 670047, Russia
erbajeva@ginst.ru
[2] Vinogradov Institute of Geochemistry, Siberian Branch RAS, Irkutsk 664033,
Russia

Abstract. Lagomorphs are an ancient group of mammals that originated in the
Paleogene of Asia. During the time of evolution a high number of taxa can be
traced. The stem group of lagomorphs inhabited Asia while tropical environmental conditions existed. Later, a gradual change towards more continental and
arid climate was driven by the Antarctic glaciation. It resulted in significant
reorganisations of paleoenvironmental and climatic conditions in Asia. Landbridges between Asia and North America and Europe, respectively, allowed
faunal exchanges, and diversification and speciation of lagomorphs occurred at
the northern continents. The earliest lagomorphs were represented by archaic
leporids and paleolagids which were gradually replaced by ochotonids, that
flourished during the Miocene and Pliocene. The diversity and abundance of
ochotonids decreased during the Pleistocene and only the genus Ochotona
survived.

Keywords: Lagomorpha · Biodiversity · Cenozoic · Eurasia

Lagomorphs are one of the ancient group of mammals known as early as the Paleogene
of Asia. At the present time the order Lagomorpha includes five families: Leporidae
Fischer, 1817; Palaeolagidae Dice, 1929; Ochotonidae Thomas, 1897; Prolagidae
Gureev, 1960, Mimotonidae Li, 1978. Only two extant families are known: Ochotonidae comprise one genus and Leporidae eleven genera. In the past they were much
more abundant and diverse than today.

The earliest record of lagomorphs is from the Paleocene of Asia [3]. At that time
the climate in Asia was warm and humid, mesophylic vegetation and broad leave
forests existed. Such conditions became favourable of earliest Paleogene mammals,
among them the stem group of lagomorphs, the Anagalida and Mimotonidae. Towards
the end of the Eocene the climate in Asia changed towards arid and continental due to
the influence of the Antarctic glaciation [24, 41]. This climate change led to the
disappearance of archaic lagomorpha at the Eocene-Oligocene transition. They were
replaced by the first forest dwelling leporide genera *Lushilagus, Shamolagus* and
Strenulagus [26, 27]. In the middle Eocene lagomorphs invaded from Asia into the
New World and adapted to new environments. From North America a number of new
genera are known, such as *Mytonolagu*s, "*Procaprolagus*", *Megalagus, Tachylagus,
Chadrolagus, Paleolagus* a.o. [8].

© Springer Nature Switzerland AG 2019
I. Bychkov and V. Voronin (Eds.): *Information Technologies
in the Research of Biodiversity,* SPEES, pp. 144–150, 2019.
https://doi.org/10.1007/978-3-030-11720-7_19

In Asia some archaic leporids persisted, and additionally the new leporid genus *Gobiolagus* and the first paleolagid genus *Desmatolagus* appeared [32]. Contrarily to Asia and North America no lagomorphs were known from Europe before the Eocene/Oligocene transition. After the "Grande Coupure" event more than 20 Asian mammal genera invaded into Europe, among them the paleolagid genus *"Desmatolagus"* [28]. During the Oligocene new endemic amphilagine genera developed in Europe, i.e. *Titanomys, Piezodus* and *Amphilagus* [29, 33, 37].

Significant reorganization of paleoenvironments occured in Asia during the Oligocene. In the Northern Hemisphere an arid zone developed from the Caspian Sea across south Kazakhstan as far as Mongolia [1]. The climate changed towards arid and continental conditions, open landscapes developed and forested environments decreased. The decrease of Leporidae evidences this development. The Eocene leporid genus *Gobiolagus* was replaced by *Ordolagus,* but the dominant Lagomorphs became the highly diversified palaeolagid *Desmatolagus* [7, 18, 19]. The earliest record of *Desmatolagus* species (D. *vetustus* and D. *robustus*) in Asia are known from the late Eocene Ardyn Obo fauna of Inner Mongolia, China [32]. In Mongolia desmatolagines are known since the early Oligocene and they became rather abundant, occurring in high individual and species numbers. They were distributed across China, Kazakhstan and Mongolia up to the end of the Oligocene. In Mongolia they are known from more than 30 locations of the Hsanda Gol, Tatal Gol and Taatsin Gol regions in the Valley of Lakes [9–11, 25]. The most numerous and characteristic forms of the early Oligocene are: *Desmatolagus gobiensis, D. orlovi, D. youngi, D. robustus.* During the late Oligocene desmatolagins flourished, some new taxa appeared, i.e. *D. shargaltensis, D. chinensis* and *D. simplex* and commonly they became the dominant group in small mammal faunas. However, at the end of Oligocene they disappear completely. Probably around the early late Oligocene transition paleoenvironment was characterized by cool condition and increasing of aridity that led to the development of the open landscapes. That phenomenon is indicated by the first appearance of steppe dwelling lagomorphs of the family Ochotonidae in the Oligocene. These earliest ochotonids were the genera *Sinolagomys* and *Bohlinotona,* although *Desmatolagus* species still dominated at that time. *Bohlinotona* occurrences were rare, and only two species are known. The evolution of this genus started with the more archaic species *B. pusilla,* which had rooted cheek teeth. The second, more advanced species *B. mongolica* n.sp. with rootless teeth developed and survived to the end of the Oligocene. The genus *Sinolagomys* developed a relatively high variety of the species both in Mongolia and Northern China—*Sinolagomys tatalgolicus, S. kansuensis, S. major, S. ulungurensis, S. gracilis, S. badamae* [4, 21]. The first two species are the most archaic ones, having rooted teeth. All other taxa are more derived and show rootless cheek teeth. The loss of tooth roots in *Sinolagomys* indicates adaptation to feeding on grass. The tendency towards rootless teeth and increasing hypsodonty of teeth evidences climatic and palaeoenvironmental changes towards semi-arid steppe environments. *Sinolagomys kansuensis* was the dominant species in late Oligocene lagomorph faunas, and rootless *S. ulungurensis* in the early Miocene.

During the Oligocene-Miocene transition again significant reorganisations of paleoenvironment and biota of Eurasia occurred. Cooling and increasing aridity led to expansion of the open landscapes. At the end of Oligocene desmatolagins gradually vanished and disappeared, *Bohlinotona* disappeared completely. However,

Sinolagomys survived the Oligocene-Miocene transition and survived to the end of the Early Miocene (*Sinolagomys ulungurensis, S. pachygnathus*).

At the beginning of the Early Miocene intensive faunal exchange between Asia and other continents happened. Asian sinolagomyine taxa invaded Africa (genera *Austrolagomys, Kenyialagomys*), Europe (genus *Heterolagus*) and migrated to North America (genus *Oreolagus*). Landbridges between continents served as migration pathways for mammalian faunas, especially for lagomorphs. Thus, the European endemic genus *Amphilagus* diversified and extended its distribution area eastwards as far as Mongolia, across Ukraine and Kazakhstan [13]. Moreover, the high diversification of ochotonids at that time can be traced through Europe and Asia. In the European faunas the genera *Amphilagus, Piezodus* and *Prolagus* continued to exist, moreover the new genera *Eurolagus, Gimnezicolagus, Marcuinonys, Lagopsis* and *Albertona* appeared. Asian faunas contained also diverse abundant taxa of the genera *Alloptox* and *Bellatona*.

The Middle Miocene was a favourable period for evolutionary development of ochotonids and amphilagins. The genus *Amphilagus* continued to occupy vast territories of Eurasia through Siberia and distributed to Japan [20, 38]. In contrast, Asian ochotonids distributed to the Europe. So, the genus *Alloptox* became the flourishing group, it distributed widely from Japan and China through Mongolia and Kazakhstan far to the west, Asia Minor (Turkey) and Hungary [2, 35, 39, 40]. The genus *Bellatona* diversified as well, but it was distributed mainly in Asia—in China, Mongolia and Kazakhstan [6, 12, 34, 42]. In Europe the advanced genus *Lagopsis* diversified successfully [5], the other genera in contrast to it were not numerous both in quantity and numbers.

During the late Miocene the climate of the Northern Hemisphere became gradually drier and cooler. The average annual temperature continued to decrease. Open landscapes with true steppes were widespread. They became main habitat for different ochotonids. New taxa of the genera *Proochotona* and *Paludotona* appeared in the Europe, and the genera *Bellatonoides, Ochotonoma, Ochotonoides* and *Ochotona* appeared in Asia, they diversified and distributed to Europe and Asia Minor in the west [17, 35]. The high taxonomic diversity of the genus *Ochotona* is evident in Asia: *O. lagrelli, O. minor, O. tedfordi, O. chowmincheni, O. guizhongensis* a.o. At the late Miocene-Early Pliocene transition, the genus *Ochotona* invaded the New World (*O. spanglei*) [36]. In turn, leporids of North America, differentiated during the early and middle Miocene, dispersed to northern Asia in the late Miocene. They spread apparently widely throughout northern Eurasia: the genera *Tsaganolagus, Trischizolagus, Hypolagus, Alilepus* are known in Asia, and the latter three genera are known as well in the Europe [22].

Further gradual cooling in the Northern latitudes and intensive orogenic processes resulted in the prominent environmental reorganisations of Eurasia during the Pliocene. Most of the Miocene ochotonid taxa disappeared completely, however, the genera *Ochotonoides* and *Ochotona* survived. During the Pliocene peculiar lagomorphs of the genus *Prolagus* continued to exist. It contained more than 25 species occurring mostly in the Europe, however, few of them are known in northern Africa and Asia Minor also. The stratigraphic range of this genus is the early Miocene through the latest Holocene.

During the Pliocene diverse taxa of the genera *Ochotonoides* and *Ochotona* occupied Eurasia. In Eastern Europe the new genus *Pliolagomys* appeared and quickly distributed eastwards. *Pliolagomys* flourished through the Pliocene and it occupied vast plain territories in Eurasia, from Romania, Moldavia and Ukraine towards west, and through western Siberia and Kazakhstan as far as Lake Baikal in eastern Siberia. Diverse and abundant taxa of the genus *Ochotonoides* were widely distributed in Asia from China across Transbaikalia and Mongolia to Kazakhstan in the west. They disappeared completely at the end of Early Pleistocene. Only the genus *Ochotona* continued to exist up to modern times.

Although *Ochotona* first appeared in the late Miocene, its main feature of skull and structure of the teeth remained unchanged to the recent times. During the Pliocene *Ochotona* flourished mostly in Asia, and in the early Pliocene (MN 14) it invaded into Europe (Maritsa). We suppose that the late Pliocene environments were favourable for pikas, thus, a high number of taxa developed and distributed all over Eurasia. There, a huge number of species existed, differing in size, and probably adapted to different habitats: *Ochotona antiqua, O. pseudopusilla, O. ursui, O. galatica, O. valerotae, O. polonica, O. dehmi, O. horaceki, O. kormosi* a.o. in the Europe, and *O. gudrunae, O. gromovi, O. tologoica, O. intermedia, O. plicodenta, O. lingtaica, O. zasuchini, O. zazhigini, O. zhangi* a.o. in Asia [15]. A few extinct ochotonid taxa are known in North America, i.e. *Ochotona spanglei, O. whartoni, O.* cf. *princeps*. The latter was widely distributed in the great intermountain basins in the west during the early and middle Pleistocene [30, 31]. The distribution area of *O. whartoni* was restricted to the western Arctic, Alaska [23]. However, this species had a much wider distribution area, it was distributed to northeast Asia, Kolyma lowland [16]. Moreover, a large *Ochotona,* close to *O. whartoni,* was evidenced in the early Pleistocene fauna of the Zayarsk locality, which is located in the northern part of Eastern Siberia, far to the west of Alaska. This taxon is associated with an advanced latest *Mimomys,* an archaic type of *Allophaiomys* and the earliest *Lemmus* (unpublished data).

Leporids in Pliocene of Eurasia became rather diverse, but they were not numerous, they were represented by the genera *Hypolagus, Alilepus, Pliopentalagus, Trischizolagus, Nuralagus, Oryctolagus* a.o. [14].

Towards the end of Pleistocene the genus *Ochotona* decreased both in abundance and quantity. Gradually they vanished in Eurasia and reduced their area of distributions probably due to the appearance of trophic and habitat competitors, namely another typical herbivorous—different arvicolids of the Microtinae subfamily. It is possible to suppose that due to the high explosive radiations, the diverse arvicolid taxa appeared, they could adapted to ochotonid' habitats—the variety landscapes—steppes, meadows, forest steppes a.o. Probably arvicolids occupied the favourable nature of ochotonids as a result the latter have been pushed out of their habitat and they were replaced by abundant variety of arvicolids. Ochotonids decreased both in quantity and numbers, few ochotonid species are known at the late Pleistocene faunas of Eurasia. They are fossil ancestors of living pikas—*Ochotona* cf. *alpina, O.* cf. *hyperborea, O. daurica, O.*cf. *rufescens* a.o. recorded in Asia. However, during the late Pleistocene the small species *Ochotona pusilla* occupied the vast plain territories of southern England, eastern Spain, Holland, France, Italy and Greece in the west, throughout the temperate zone of Eurasia, to the Prebaikal region of East Siberia. At the present time this species exists only in Northern Kazakhstan and in the restricted area of steppe in Upper Volga river (Russia) [12].

Totally, more than 40 extinct and 28 extant ochotonid species are known in the World. Presently, modern ochotonids (15) occupy mostly the territory of Asia, two taxa are known in North America and one species is restricted to easternmost Europe.

Leporids in Asia reduced their variety and inhabited south-eastern territory. In Europe all genera disappeared, except for the genera *Lepus* and *Oryctolagus*. At present 11 leporid genera are known from the world.

Acknowledgements. The study was supported by grants RNF, Nr 16-17-10079 (Russian Science Foundation) and FWF, Nr P-10505-GEO (Austrian Science Foundation).

References

1. Akhmetiev, M.A.: The climate of the epoch in context of the significant changes in Biosphere. Transactions of the Geological Institute RAS, issue 350, Moscow, Nauka Press (2004). (in Russian)
2. Angelone, Ch., Hir, J.: Alloptox katinkae sp. nov. (Lagomorpha, Ochotonidae), westernmost Eurasian record of the genus from the early Middle Miocene vertebrate fauna of Litke 2 (N Hungary). Neues Jahrbuch fuer Geologie und Palaeontologie, Abhandlungen **264**(1), 1–10 (2012)
3. Asher, R.J., et al.: Stem Lagomorpha and the antiquity of Glires. Science **307**, 1091–1094 (2005)
4. Bohlin, B. Oberoligocene Säugetiere aus dem Shargaltein-Tal.—Sino-Swedish Expedition. VI, 2.—Palaeont. Sinica, New Ser. C, 3 (Whole Ser. 107), pp. 1– 66 (1937)
5. Bucher, H.: Etude des genres Marcuinomys Lavocat et Lagopsis Schlosser (Lagomorpha, Mammalia) du Miocene inferieur et moyen de France. Implications biostratigraphiques et phylogenetiques. Bull. Mus. Natl. Histoire naturales. Ser. 4, Sect. C **1/2**, 43–74 (1982)
6. Dawson, M.R.: On two ochotonids (Mammalia, Lagomorpha) from the later Tertiary of Inner Mongolia. Am. Mus. Novit. **2061**, 1–15 (1961)
7. Dawson, M.R. Lagomorph history and the stratigraphic record. In: Essays in Paleontology & Stratigraphy. Raymond C. Moore Commemorative, pp. 287–316 (1967)
8. Dawson, M.R.: Lagomorpha. In: Janis, Ch.M., Gunnell, G.F., Uhen, M.D. (eds.) Evolution of Tertiary Mammals of North America, vol. 2, pp. 293–310. Cambridge University Press (2008)
9. Daxner-Hoeck, G., Badamgarav, D.: Geological and stratigraphical setting. In: Daxner-Hoeck, G. (ed.) Oligocene-Miocene vertebrates from the Valley of Lakes (Central Mongolia): Morphology, Phylogenetic and Stratigraphic Implications. Annalen des Naturhistorischen museums in Wien, Serie A, vol. 108, pp. 1–24 (2007)
10. Daxner-Hoeck, G., et al.: Cenozoic stratigraphy based on a sediment-basalt association in Central Mongolia as requirement for correlation across Central Asia. Mémoires et Travaux de l'Institut de Montpellier, E.P.H.E. **21**, 163–176 (1997)
11. Daxner-Hoeck, G. et al.: Oligocene stratigraphy across the eocene and miocene boundaries in the Valley of Lakes (Mongolia). In: Daxner-Hoeck, G., Goelich, U. (eds.) The Valley of Lakes in Mongolia, a Key Area of Cenozoic Mammal Evolution and Styrastigraphy vol. 97, no. 1, pp. 111–218 (2017). Palaeobiodiversity and Palaeoenvironments

12. Erbajeva, M.A.: Cenozoic pikas (Taxonomy, Systematics and Phylogeny). Nauka, Moscow (1988). (in Russian)
13. Erbajeva, M.A.: New species of Amphilagus (Lagomorpha, Mammalia) from the Miocene of the Valley of Lakes, Central Mongolia. Paleontol. J. **47**(3), 311–320 (2013)
14. Erbajeva, M.A.: The evolution and biodiversity of cenozoic holarctic lagomorphs in context of global climate change. Sci. J. Vestn. BSC Siberian Branch RAS. **1**(13), 134–141 (2014). (in Russian)
15. Erbajeva, M.A.: The ochotonids of Eurasia: biochronology and taxonomic diversity. Biol. Bull. **43**(7), 729–735 (2016)
16. Erbajeva, M.A., Belolyubski, I.N.: First discovery of Ochotona whartoni in Kolyma lowland. Russ. J. Geol. Geophys. **34**(6), 142–144 (1993)
17. Erbajeva, M.A.: Lagomorpha (Mammalia): preliminary results. In: Daxner-Hoeck, G. (ed.) Oligocene-Miocene Vertebrates from the Valley of Lakes (Central Mongolia): Morphology, Phylogenetic and Stratigraphic Implications. Annalen des Naturhistorischen museums in Wien, Serie A, vol. 108, pp. 165–171 (2007)
18. Erbajeva, M.A., Daxner-Hoeck, G.: Paleogene and Neogene Lagomorphs from the Valley of Lakes Central Mongolia. Lynx **32**, 55–65 (2001)
19. Erbajeva, M.A., Daxner-Hoeck, G.: The most prominent Lagomorpha from the Oligocene and Early Miocene of Mongolia. Annalen des Naturhistorischen museums in Wien, Serie A **116**, 215–245 (2014)
20. Erbajeva, M.A., Angelone, Ch., Alexeeva, N.V.: A new species of the genus Amphilagus (Lagomorpha, Mammalia) from the Middle Miocene of south-eastern Siberia. Hist. Biol. **28** (1–2), 199–207 (2016)
21. Erbajeva, M.A., Bayarmaa, B., Daxner-Hoeck, G., Flynn, L.J.: Occurences of Sinolagomys (Lagomorpha) from the Valley of Lakes (Mongolia). In: G. Daxner-Hoeck and U. Goelich (Eds.) The Valley of Lakes in Mongolia, a key area of Cenozoic mammal evolution and styrastigraphy, vol. 97, pp. 11–24 (2017). Palaeobiodiversity and Palaeoenvironments
22. Flynn, L., Winkler, A.J., Erbajeva, M.A., Alexeeva, N., Anders, U., et al.: The Leporid datum: a late Miocene biotic marker. Mammal Rev. **44**, 164–176 (2014)
23. Guthrie, R.D., Matthews, J.W.: The Cape Deceit fauna—early Pleistocene Mammalian assemblage from the Alaskan Arctic. Quternary Res. **1**(4), 474–510 (1971)
24. Harzhauser, M., et al.: Oligocene and early Miocene mammal biostratigraphy of the Valley of Lakes, Mongolia. In: Daxner-Hoeck, G., Goelich, U. (eds.) The Valley of Lakes in Mongolia, a Key Area of Cenozoic Mammal Evolution and Styrastigraphy, vol. 97, pp. 219–231 (2017). Palaeobiodiversity and Palaeoenvironments
25. Hoeck, V., et al.: Oligocene-Miocene sediments, fossils and basalts from the Valley of Lakes (Central Mongolia). An integrated study. Mitt. Österr. Geol. Ges. **90**, 83–125 (1999)
26. Li, C.K.: Eocene leporids of North China. Vertebrata Palasiatica **9**(1), 23–33 (1965). (in Chinese with English Summary)
27. Li, Ch., Ting, S.-Y.: The Paleogene mammals of China. Bull. Carnegie Mus. Nat. Hist. **21**, 1–93 (1983)
28. Lopez Martinez, N., Thaler, L.: Sur le plus ancient Lagomorphe europee et la "Grande Coupure" Oligocene de Stehlin. Palaeovertebrata **6**, 234–251 (1974)
29. Major, F.C.I.: On Fossil and recent Lagomorpha. Trans. Linn. Soc. Lond. Ser. 2, **7**, 433–520 (1899)
30. Mead, J., Grady, F.: Ochotona (Lagomorpha) from late Quaternary cave deposits in eastern North America. Quaternary Res. **45**, 93–101 (1996)
31. Mead, J., Erbajeva, M., Swift, S.L.: Middle Pleistocene (Irvingtonian) Ochotona (Lagomorpha, Ochotonidae) from Porcupine Cave. In: Barnosky, A.D. (ed.) Biodiversity

Response to Climate Change in the Middle Pleistocene. Porcupine Cave Fauna from Colorado, pp. 155–164. University of California Press (2004)

32. Meng, J., Hu, Y., Li, Ch.: Gobiolagus (Lagomorpha, Mammalia) from Eocene Ula Usu, Inner Mongolia, and comments on Eocene Lagomorphs of Asia. Palaeontologia Electronica **8**(1), 7A:1–23 (2005)

33. Moers, T., Kalthoff, D.: A new species of Amphilagus (Mammalia, Lagomorpha) from the Late Oligocene lake deposits of Enspel (Westerwald, Germany). Palaeobiodiversity Palaeoenvironments **90**, 83–98 (2010)

34. Qiu, Zh.: Middle Miocene Micromammalian fauna from Tunggur, Nei Mongol, Beijing, pp. 1–214 (1996)

35. Sen, S.: Lagomorpha. In: Fortelius, M., Kappelman, J., Sen, S., Bernor, R.L. (eds.) Geology and Paleontology of the Miocene Sinap Formation, Turkey, pp. 163–177. Columbia University Press, New York (2003)

36. Shotwell, J.A.: Hemphilian mammalian assemblage from north-eastern Oregon. Geol. Soc. Am. Bull. **67**(6), 717–738 (1956)

37. Tobien, H.: Zur Gebissstruktur, Systematik und Evolution der Genera Amphilagus und Titanomys (Lagomorpha, Mammalia) aus einigen Vorkommen im jungeren Tertiar Mittel- und Westeuropas. Mainz. Geowiss. Mitt. **3**, 95–214 (1974)

38. Tomida, Y, Goda, T.: First discovery of Amphilagus-like ochotonid from the Early Miocene of Japan. In: Annual Meeting Paleontological Society of Japan, Abstracts: 76 (1993)

39. Tomida, Y.: New species of Alloptox (Lagomorpha, Ochotonodae), first record of the genus in Japan, and subgeneric distinction. Paleontol. Res. **16**, 19–25 (2012)

40. Unay, E., Sen, S.: Une nouvelle espece d'Alloptox (Lagomorpha, Mammalia) dans le Tortoniennde l'Anatolie. Bull. Miner. Res. Explor. Inst. Turk. **85**, 145–149 (1976)

41. Wolfe, J.A.: Tertiary climatic fluctuations and methods of analysis of tertiary floras. Palaeogeogr. Palaeoclimatol. Palaeoecol. **9**, 27–57 (1971)

42. Zhow, X.: Miocene ochotonid (Mammalia, Lagomorpha) from Xinzhou, Shanxi. Vertebrata PalAsiatica **26**(4), 139–148 (1988). (in Chinese with English Summary)

An Instrumental Environment for Metagenomic Analysis

Evgeny Cherkashin[1,2,4(✉)], Alexey Shigarov[1,2], Fedor Malkov[1,2,4] [iD],
and Alexey Morozov[3] [iD]

[1] Irkutsk Scientific Center of SB RAS, Lermontov Str. 134, Irkutsk 664033,
Russia
eugeneaig@icc.ru
[2] Matrosov Institute for System Dynamics and Control Theory of SB RAS,
Lermontov Str. 134, Irkutsk 664033, Russia
[3] Limnological Institute of SB RAS, Ulan-Bator Str. 3, Irkutsk 664033, Russia
[4] National Research Irkutsk State Technical University, Lermontov Str. 83,
Irkutsk 664074, Russia

Abstract. Metagenomic analysis allows describing microbial community with
a previously unavailable precision, but requires considerable computing power
for solving bioinformatics problems and participation of domain specialists at
the stage of the result interpretation. This complicates the implementation of the
analysis in a broad biological practice. The development of a domain user-
friendly software environment for storage and analysis of metagenomic data has
been started. The usage of a data ow programming system for representation of
metagenomic analysis and the schema for a SQL database for storage of the
metadata are considered as units of the environment.

Keywords: Bioinformatics · Metagenomic analysis · Big Data

1 Introduction

In the last decade, thanks to the invention of next-generation sequencing
(NGS) methods and their introduction in practice of research of biological systems a
field of research of molecular genetics, namely metagenomics, has been arisen. Its basic
principle is that the object under investigation is not a separate microscopic organism,
but their communities (microbiomes). The sampled probes stand out with a total DNA
sequencing data over the whole set of genes of all microorganisms in the probe. That is,
the studied object is the microbiome as a whole, not only those organisms which can be
cultivated in laboratory conditions or identified with microscopic or microbiological
methods.

Metagenomics allows us to describe a significant number of new groups on all
taxonomic levels, broadening the field of view of the world science. A characteristic
example is the recently discovered group CPR (*candidate phyla radiation*). No CPR is
isolated in a culture at the moment. According to genomic data, its representatives
differ in the set of ribosomal proteins, the absence of certain key metabolic pathways
and the presence of self-splicing introns in genes 16S rRNA [1]. Phylogenetic analysis

© Springer Nature Switzerland AG 2019
I. Bychkov and V. Voronin (Eds.): *Information Technologies
in the Research of Biodiversity*, SPEES, pp. 151–158, 2019.
https://doi.org/10.1007/978-3-030-11720-7_20

indicates that this group is a sister to all other bacteria, and the level of divergence is not inferior to bacteria, not to mention the eukaryotes [2].

There are two main types of metagenomic studies. The first one, which is simpler, is called *analysis of the amplicons*. In this case a specific taxonomic marker is amplified and sequenced. The marker is universal for the studied species. Usually, the sequence of the small subunit of ribosomal RNA is used as the marker, as this gene is widely used in phylogenetics. The gene is available in numerous reference sequences. For example, the release 128 of widely used in amplicon analysis SILVA database [3] contains 645 151 unique rRNA. The reads obtained from the DNA sequences extracted from the sample under investigation are compared to the sequences in databases, attributing them to a particular taxon of a taxonomic level, obtaining information about the diversity of the microbiome in the studied environment.

The second approach is known as *metagenomic shotgun method*. It is based on sequencing the whole DNA sample instead of the specific locus. With sufficient coverage, this approach allows describing the taxonomic composition of the community, as well as the genes of functional or structural proteins presented in the representatives of the community, including viral ones [4]. On the basis of metagenomic data, metabolic interactions in individual microbiomes can be determined using the databases ePGDBs (environmental pathway/genome databases) [5]. In several works, full genomes of individual species were isolated from metagenomic dataset reads [6].

In recent years the amplicon analysis was applied in microbiome studies for different environments of lake Baikal. The researchers of the Limnological Institute of SB RAS described the under-ice bacterial communities associated with blooms of diatoms [7] and bacteria in photic layer during spring [8]. Bacteria inhabiting the Baikal sponges were studied as well [9]. Finally, the bacterial communities of bottom sediments in the areas of hydrocarbon yields [10, 11] were investigated.

In order to carry out the metagenomic studies the significant computational resources and bioinformatics skills are required for data processing and interpretation. The software used for analysis of amplicons includes various library modules of sequence processing, for example, Mothur [12], USearch [13], statistical packages and development environments of data mining algorithms, e.g., R (https://www.r-project. org). In order to carry out the studies of metagenomic data, the specialists are required to be able scripting the command shell of an operating system (Linux, Windows), running programs in a distributed computing environment and cluster computing systems, and programming with general-purpose languages, usually R or Python.

Another important problem is the organization of a centralized data storage and providing the efficient regulated access to the data for the users. At the moment the staff of LIN SB RAS conducted numerous amplicon research of the different ecotopes of lake Baikal, the data were collected for several years. There are no strict rules of the storage policy of input, intermediate data and the obtained results. Comparison and integration of data from different studies is also complicated due to its heterogeneity, resulting from the use of various software. The implementation of a system for storing input data, metadata, and results of metagenomic studies in a unified form will simplify the integration of results from different studies and the comparative analysis.

The goal of this study is a software environment development for supporting the processes of new-generation sequencing with organizational, informational and computational resources.

2 The Domain Analysis

Domain analysis showed that the problems solved in the bioinformational part of metagenomic analysis, together with NGS itself, are well represented within the paradigm of Big Data. At the moment, the scientific community developed data formats for representation and storage of metagenomic information, algorithms and software modules including distributed and parallel implementations on cluster computing systems providing different stages of data analysis.

The solution of the problems within the Big Data paradigm requires the biologist to have software development skills to be a professional programmer in bioinformatics. In order to carry on the analysis of each probe, biologist is to construct and execute a separate program script or perform stage-by-stage execution manually to control each step's results quality. This approach significantly slows down the process of obtaining the final result.

The proposed organization of studies is based on the creation of an information-computational environment that allows one to design and execute scenarios, giving the input data in various formats from various sources, e.g., files, databases, servers of metagenomic information. The environment must also support a cloud storage for intermediate data and the obtained results. A collaborative project of LIN SB RAS and ISDCT SB RAS is devoted to the construction of the environment for the research support. The following problems are to be solved within the project.

1. The subject area and its functional modeling. The classes of functions (problems) are being recognized and presented in the form of software modules. Modules form scripts of problem solving, network graphs of modules connected by data transmission.
2. Metadata descriptions of the modules and structures of input and output data. At this stage, it is necessary to deal with the problem of integration with external information and computational resources. In this case, the standards and standard means of data modeling like ontologies are of critical usage.
3. Decomposition of the input/output data formats and implementation of subsystems of their transformation, accumulation, storage and effective (according to the criteria of time and computational complexity) regulated access.
4. Construction of virtual executional environments and software interfaces for modules, whose source code is inaccessible due to the lack of the source code or licensing restrictions.
5. Development a customized user interface for high level control of the scenario executions. At this stage, a visual programming with the user interface for script development and execution is required to provide flexibility for managing computational processes by domain specialists.

6. Development of subsystems of visualization and interpretation of obtained results, including the modules for interpretation of the process of metagenomic analysis.

3   Dataflow Representation of the Domain

A popular approach to the representation of the computational process is dataflow programming [14]. The data flow programs are constructed as a combination of the executable modules. The modules receive input data, process it, and produce output. The approach is being developed since the 1970-ies.

An example of usage of the script construction system under development is shown in Fig. 1. The figure shows an example of an initial stage of a computing process of analysis of the amplicons.

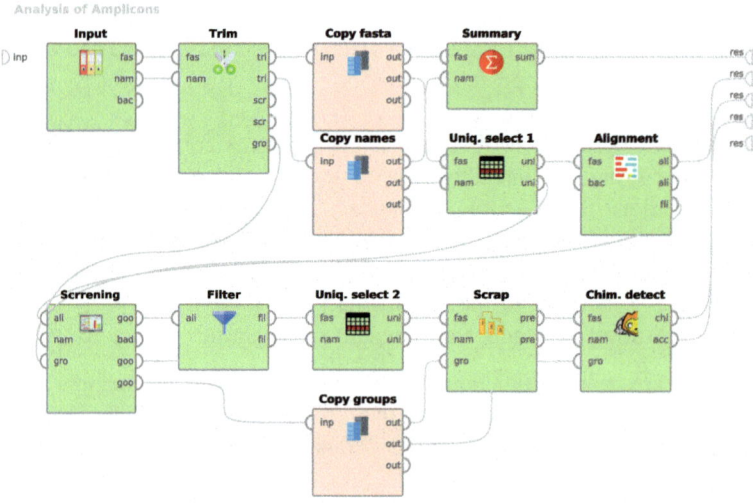

Fig. 1. The initial stage of the metagenomic analysis represented as a data flow.

The presented script was constructed by means of the software package Rapidminer Studio (https://rapidminer.com/) supplemented by our extension module for description of the amplicon analysis stages. The scenario includes the following operations:

- definition of a research project as a set of input files containing the sequencing data in a directory (module "Input");
- trimming reads (module "Trim");
- the module "Summary" is used for visual quality analysis of the results of the previous steps;
- the reduction of the volume of input by the removal of insignificant information, for example, overlapping sequences (modules, "Uniq. select …");
- alignment of sequences to the reference database (module "Alignment");

- filtering sequences according to specified criteria (module "Screening");
- removing alignment columns based on specified criteria, for example, empty columns (module "Filter");
- removal of sequences containing sequencing error (the module "Scrap");
- detection of chimeras (module "Chim. detection), etc.

The diagram shows service modules of RapidMiner Studio, which are necessary to distribute the same type of information between modules (e.g., "Copy groups"). The necessity of introduction of such modules is a feature of Rapidminer Studio; it supposes that in a general case the modules make changes in the data under processing without copying it.

Each module receives file names as input and creates new file set as the result. The operation of the module depends on the parameters specified by the user via user interface of each module. The results of the script are sent to output ports and displayed by the Rapidminer Studio visualization subsystem in a convenient form to the user. The system supports presentation of a scenario as a new block with its input and output ports, as well as a cloud storage and execution of scripts, creating distributed computing environment. Rich feature set of Rapidminer Studio and various services provided by its developers were the main reason for choosing this system as a development environment for informational-computation resources of the project.

4   The Database Supporting Metagenomic Analyses

Assessment of world experience of organization of scientific research in the field of Data Science showed that the use of cloud technologies is a necessary basis for the interaction organization of the researchers. A specialized data storage should be a unit of the environment to ensure effective user access and computing processes to the data of research.

Database for microbiome-based metagenomic analysis data (Fig. 2) provides the storage facility on all the stages of the microbiome studies from the probe sampling to the publication of the scientific meaningful results. The scheme in the Fig. 2 represents database structure as an ER-diagram. The scheme contains data about sampling, analysis of physicochemical and biological parameters of the probes, the sequencing results, the applied equipment and software, taxonomic databases, methods of the analysis of the collected material, publications of the obtained results and the participated researchers. It also allows us to store the processing scripts of analysis of metagenomic data, including software tools, commands, and configuration files. The latter function allows one to save the state of the computational process and restart it from the specified point.

The model is implemented by means of Django framework (https://www.djangoproject.com). The framework supports automatic definition of rational table structures representing many-to-many relations, and generation a customizable interface for the administrative panel, allowing testing the developed model. These same tools are used for the implementation of the project web site.

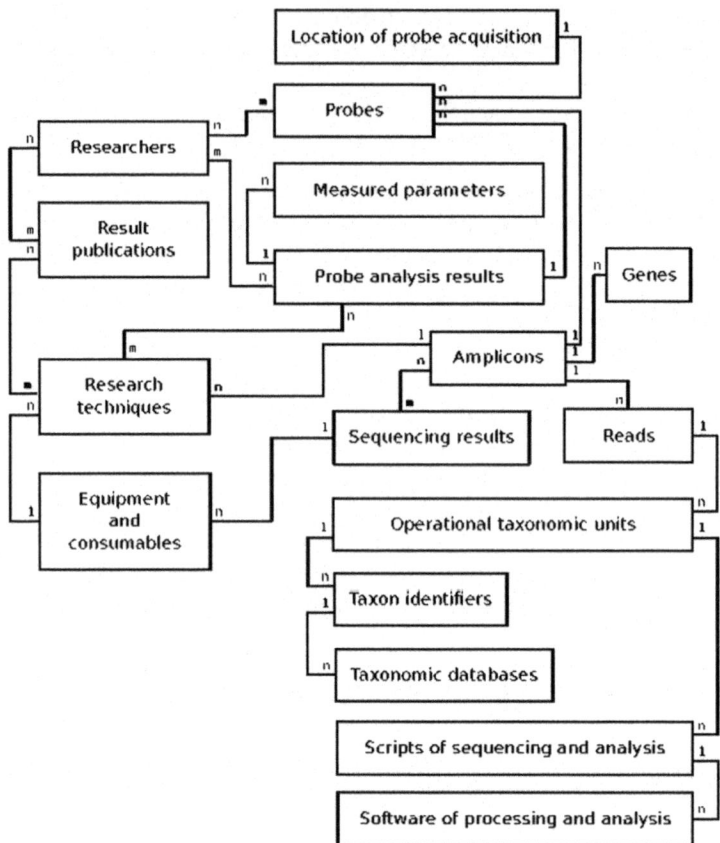

Fig. 2. A general schema representation of the database for microbiome research based on metagenomic analysis.

Cloud-based storage and dedicated storage of the metagenomic data will allow one to create online services for joint processing of sequencing data from different studies by specialized software and to publish information in the Internet. In order to achieve this goal, the following must be carried out:

- a software interface implementation for data access;
- filling in the database with information collected and processed as a result of studies of the microbiome of lake Baikal in 2009–2015;
- realization of the scenario design and execution to support the metagenomic data analysis in a distributed computing environment.

5 Conclusion

Modern problems of development of a distributed software environment for the implementation of organizational, informational and computational resources for scientific microbiological studies based on metagenome analysis are presented in the article. A generalized domain model of system-level is conducted, as well as the requirements are stated to the development environment and problems to be solved. A computational model of the process of analysis of the amplicons is being constructed and implemented. Aspect of informational supply of the computational process is represented by realization of the problem of cloud storage for computing processes (scenarios of metagenomic data processing), as well as by construction of a database for storing input and intermediate data, results of the scenarios execution. The database is used as a basis of an information portal construction for processing metagenomic data and presenting the results of scientific community.

Acknowledgment. This work was supported by the project of Irkutsk scientific center, SB RAS No. 4.2.

References

1. Brown, C.T., Hug, L.A., Thomas, C.B., et al.: Unusual biology across the group comprising more than 15% of domain Bacteria. Nature **523**, 208–211.1 (2015)
2. Hug, L.A., Brett, J.B., Anantharaman, K., et al.: A new view of the tree of life. Nat. Microbiol. **1**, 16048 (2016)
3. Quast, C., Pruesse, E., Yilmaz, P., et al.: The SILVA ribosomal RNA gene database project: improved data processing and web-based tool. Nucleic Acids Res. **41**, 590–596 (2013)
4. Paez-Espino, D., Eloe-Fadrosh, E.A., Pavlopoulos, G.A., et al.: Uncovering earth's virome. Nature **536**, 425–430 (2016)
5. Hanson, N.W., Konwar, K.M., Wu, S.J., Hallam, S.J.: Introduction to the analysis of environmental sequence information using metapathways. Comput. Methods Next Gener. Seq. Data Anal. 25–56 (2016)
6. Iverson, V., Morris, P.M., Frazar, C.D.: Untangling genomes from metagenomes: revealing an uncultured class of marine Euryarchaeota. Science **335**, 587–590 (2012)
7. Bashenkhaeva, M.V., Zakharova, Y.R., Petrova, D.P.: Sub-Ice microalgal and bacterial communities in freshwater Lake Baikal, Russia. Environ. Microbiol. **70**(3), 751–765
8. Mikhailov, S., Zakharov, Y.R., Galachyants, Y.P., et al.: On the uniformity in taxonomic composition of bacterial communities in the photic layer of the three basins of lake Baikal, which differ in composition and abundance of spring phytoplankton. Rep. Acad. Sci. **465**(5), 620–626 (2015)
9. Gladkikh, A.C., Kalyuzhnaya, O.V., Belykh, O.I., Ahn, T.S., Parfenova, V.V.: Analysis of the bacterial community of two endemic sponges from lake Baikal. Microbiology **83**(6), 682–693 (2014)
10. Zemskaya, T.I., Lomakina, A.V., Mamaeva, E.V., et al.: Bacterial communities in sediments of Lake Baikal from areas with oil and gas discharge. Aquat. Microb. Ecol. **75**, 95–109 (2015)

11. Bukin, S.V., Pavlova, O.N., Manakov, A.Y., et al.: The ability of microbial community of Lake Baikal bottom sediments associated with gas discharge to carry out the transformation of organic matter under thermobaric conditions. Front. Microbiol. **7**, 690 (2016)
12. Schloss, P.D., et al.: Introducing Mothur: open-source, platform-independent, community-supported software for describing and comparing microbial communities. Appl. Environ. Microbiol. **75**(23), 7537–7541 (2009)
13. Edgar, R.C.: Search and clustering orders of magnitude faster than BLAST. Bioinformatics **26**(19), 2460–2461 (2010)
14. Johnston, W.M., Hanna, J.R.P., Millar, R.J.: Advances in dataflow programming languages. ACM Comput. Surv. **36**, 1–34 (2004)

Experience in the Use of GIS Tools in Plant Systematics and Conservation

M. Olonova[(⊠)], D. Feoktistov, and T. Vysokikh

Tomsk State University, Lenin Av., 36, Tomsk 634050, Russia
olonova@list.ru

Abstract. The ecological niche and the range are, together with the morphotype, a sufficient characteristic of the species. In turn, each species has its own unique ecological niche, ecologo-climatic characteristics to be its important component. The MaxEnt algorithm allows not only to obtain a species distribution model, but also to evaluate it, using the AUC. It is also possible to estimate the impact of each bioclimatic variable on the model. The opportunities provided by the climate modeling programs Bioclim and MaxEnt, as well as the SDMtoolkits and ENMTools applications, were implemented in our study of bluegrasses (*Poa* L.). Using freely available climatic data and the data of occurrence, the bioclimatic profiles of morphologically similar *Poa palustris* L., *P. nemoralis* L. and populations, combining the characters of their both, treated as Aggr. *P. intricata* Wien., were revealed, using the Bioclim program, implemented in DIVA-GIS. The models for their potential distribution in the current climate, in Pleistocene maximal glaciation, in interglacial and the Middle Holocene were reconstructed with MaxEnt and applications. The comparison methods—niche-identity test (I-test), and background test (B-test), implemented in the ENMTools program, make it possible to compare the obtained ecologo-climatical niches. We compared the ecologo-climatical niches of three similar species, and I-test has revealed their differences at a statistically significant level. The models of potential species distribution, constructed on the basis of ecologo-climatical niches can be used not only for paleogeographical reconstructions, but also are of a great practical value.

Keywords: Ecologo-climatic modeling · MaxEnt · *Poa L.*

The ecological niche of the species, formed as a result of adaptations to different environmental conditions, is its inherent feature, like the morphological characteristics and range. The identification of ecological niche, its divergence and other changes is of no less importance for the study of speciation and phylogenetic reconstructions than the study of changes in morphological characters. Besides of having the great importance for study of evolution, these data are very significant for predicting of invasive species distribution, conservation biology and for introduction of plants [1–4]. The problem of a correct and objective assessment of the most important environmental factors, as well as the determination of the ecological status of species, is becoming increasingly important in modern ecology. It is known that at the present time there are many concepts of the ecological niche, but the predominant model is the concept of [5] which is treated as an n-dimensional volume within the axes of ecological factors. The every

© Springer Nature Switzerland AG 2019
I. Bychkov and V. Voronin (Eds.): *Information Technologies in the Research of Biodiversity,* SPEES, pp. 159–168, 2019.
https://doi.org/10.1007/978-3-030-11720-7_21

species has its own range within each factor, where it can exist [6, 7]. If you draw projections from the extreme points of the ranges of each factor axis, you get an n-dimensional figure, where n is the number of significant environmental factors. Ecological and climatic characteristics of the species are an important component of its ecological niche and give a general idea of the climatic needs of the species over large areas.

The introduction of geoinformation technologies into practice of environmental studies gave a new powerful impetus for the development of this research. These methods, initially focused on purely practical needs of geodesists, geographers and geologists, have now found wide application in the ecology of plants and are rapidly developing [8–10].

On the one hand the environmental modeling methods, based on GIS technologies, make it possible to identify the bioclimatic profile of the species, the part of the ecological niche that is determined by such biologically significant climatic parameters as mean annual temperature, mean annual precipitation, average temperature of the hottest quarter, average temperature of the driest quarter, seasonality of precipitation. On the other hand, the data obtained are the basis for construction of a distribution model, identifying areas of suitable climate conditions for growth [11–13].

These models can be compiled not only for the present time, but also for past geological periods, and even provide forecast maps for their possible future distribution in accordance with certain scenarios of climate change [10, 14].

Nevertheless, it should be emphasized, that it is only a simulation of the probabilistic distribution of climatic conditions favorable for the growth of a species, whereas the success of their introduction into plant communities and anchoring in them depends to a large extent on other causes - the competitiveness of the species, its biological features, interrelations of the components of the community.

There are several methods of bioclimatic modeling at present. As a rule, they require the geographical coordinates of not only those points where the species is present (was collected or registered), but also points where it is guaranteed to be missing. Of course, this approach is applicable only to well-studied territories, and is completely inappropriate for research in Siberia and Central Asia. Methods that are satisfied with data only about the presence of species are not so many. The most popular currently is the MAXEnt algorithm [15, 16] implemented and visualized in the Diva-GIS program [8, 17].

This application is based on the identification of the climatic niche of the species under study, which is established by combining the data of the geographical distribution of species (geographical coordinates) and the climatic characteristics of these points. The Worldclim database [18] contains 19 biologically significant climate variables with different spatial resolution, and it is freely available on the internet (BIO1 = Annual mean temperature BIO2 = Mean diurnal range (max temp—min temp) (monthly average) BIO3 = Isothermality (BIO1/BIO7) * 100 BIO4 = Temperature Seasonality (Coefficient of Variation) BIO5 = Max Temperature of Warmest Month BIO6 = Min Temperature of Coldest Period BIO7 = Temperature Annual Range (BIO5-BIO6) BIO8 = Mean Temperature of Wettest Quarter BIO9 = Mean Temperature of Driest Quarter BIO10 = Mean Temperature of Warmest Quarter BIO11 = Mean Temperature of Coldest Quarter BIO12 = Annual Precipitation

BIO13 = Precipitation of Wettest Month BIO14 = Precipitation of Driest Month BIO15 = Precipitation Seasonality (Coefficient of Variation) BIO16 = Precipitation of Wettest Quarter BIO17 = Precipitation of Driest Quarter BIO18 = Precipitation of Warmest Quarter BIO19 = Precipitation of Coldest Quarter). If desired, and with the availability of data, you can add other indicators—for example, such as slope, salinity, chemical composition of soils. The DIVA-GIS program allows you to determine the bioclimatic parameters of each point where a view was recorded, then, build various histograms, and conduct multivariate analyses of data.

The BIOCLIM program [14], which is also implemented in DIVA-GIS, allows one to evaluate the view for each bioclimatic parameter separately and compare it with different species. It visualizes the ecological niche occupied by a species, in two ways: in the form of a histogram, and an "envelope". The histogram shows the frequencies of the various climatic parameters observed in the species in a given area [10]. The abscissa indicates the strength of the factor, and the ordinate shows the frequencies (Fig. 1). Using these options, 19 bioclimatic characteristics of *Poa palustris, P. nemoralis* and *P. compressa* were obtained for research on their ecological niches. Some of them are represented for *P. palustris* in Fig. 1.

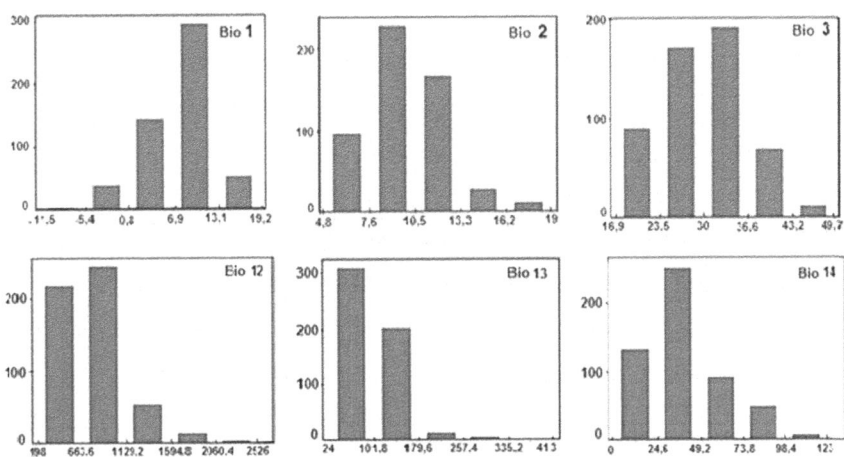

Fig. 1. Some bioclimatic characteristics of *Poa palustris* ecological niche.

In addition, it is possible to present the histograms for different species on the same plot to compare their indicators (Fig. 2). "Envelope" is a two-dimensional view of a species niche based on two climatic parameters. The revealing of the ecological niche of the species is based on the identification of the climatic characteristics of the points where the species was found and the construction on this basis a rectangle representing the coordinates of climatic variables that limit the growth range of the species. The dashed line inside the "envelope" indicates the area where values fall within the range of 5 to 95 percentile.

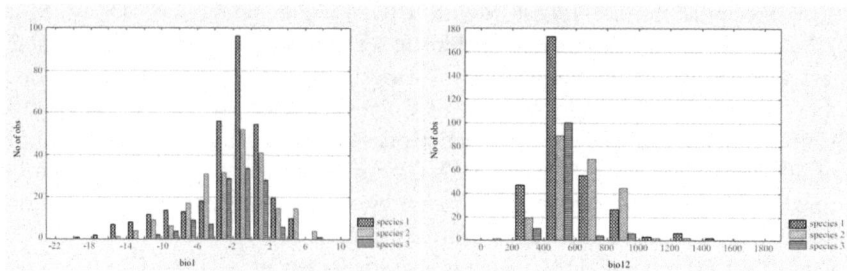

Fig. 2. Comparison of two bioclimatic characteristics of *Poa palustris* (1), *P. nemoralis* (2) and *P. intricata* (3) in Asian Russia.

Figure 3 shows the scatter plot of *Poa palustris* points in the territory of Asian Russia covering the coordinates of the annual average temperature and average annual precipitation [19].

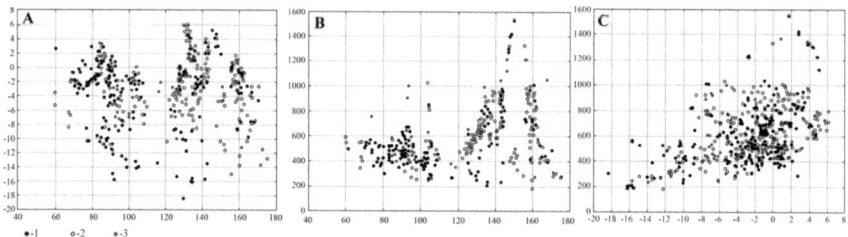

Fig. 3. Ecologo-climatic distribution of *Poa palustris* (1), *P. nemoralis* (2) and *P.intricata* (3) in Asian Russia (**A** – longitude/annual temperature, **B** – longitude/annual precipitation, **C** – annual temperature/annual precipitation).

The interesting data comparing individual bioclimatic characteristics can be displayed on two-dimensional graphs. Consideration of the graphs of pairs of precipitation-temperature points in various combinations makes it possible to better identify possible divergence of different species and adaptation to the current environmental conditions. Figure 3 shows graphs of the point scatter of three species of bluegrass (*Poa* L.) along the coordinates of the longitude and the average annual temperature, longitude of the terrain and average annual precipitation, average annual temperature and average annual precipitation.

The modern methods allow us not only to identify the ecological and climatic component of the ecological niche of species, but also to compare them, revealing a statistically significant difference between them. Currently, the niche-identity test (I-test), implemented in the ENMTools program [20, 21], which makes it possible to compare ecological niches, is widely accepted. This test allows us to evaluate hypotheses about the identity of niches and is an important tool for investigating divergence and evolution.

The I-test, used for taxa with overlapping ranges, provides a comparative analysis of ecological and climatic niches using measures I—the standardized Hellinger distance and the D—Schöner index [20].

At the same time, as Zink [22] notes, if different populations live in different climatic conditions, the I-test can have a high value, but this does not yet indicate their environmental-climatic divergence, since they can exist in different conditions due to environmental plasticity. In order to identify the ecological factors, and in this case, ecologo-climatic divergence at the genetic level, it is necessary to use a more complex B-test, which allows us to detect the presence of a genetically determined discrepancy for niches. Both tests are recommended to be carried out with all 19 variables in 100-fold repetition (100 replicates).

Figure 4 shows the result of the test for the identity of niches of two species of bluegrass—*P. palustris* and *P. nemoralis*. As a result of pairwise comparison of ecological and climatic niches constructed on the basis of all 19 biologically significant climatic variables for these species, the histograms with total information of 100 replicates were obtained. On the graphs (Fig. 4) the arrows indicate the position of the value, indicating overlapping of ecological-climatic niches between species. The diagram on the right illustrates the distribution of overlapping in replicates. On all graphs, the arrows stand at a considerable distance from the diagrams. This indicates that the null hypothesis about the identity of niche models should be rejected (Fig. 5).

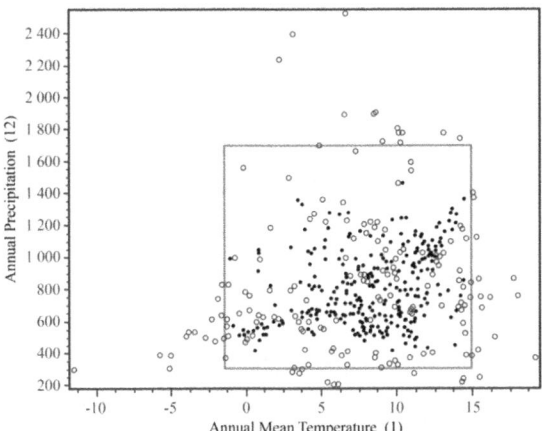

Fig. 4. Ecologo-climatic "envelope", obtained for *Poa palustris* 2 climatic variable.

The second important component of environmental and climate research is to obtain a model for the likely distributed of the species in accordance with its climatic needs. To obtain and evaluate the model, the sample is divided into 2 parts training and testing samplings. The training sampling serves to identify the ecological and climatic profile, and the test samplings is for verifying the quality of the obtained model. The resulting model is then projected onto an electronic map of the region under study.

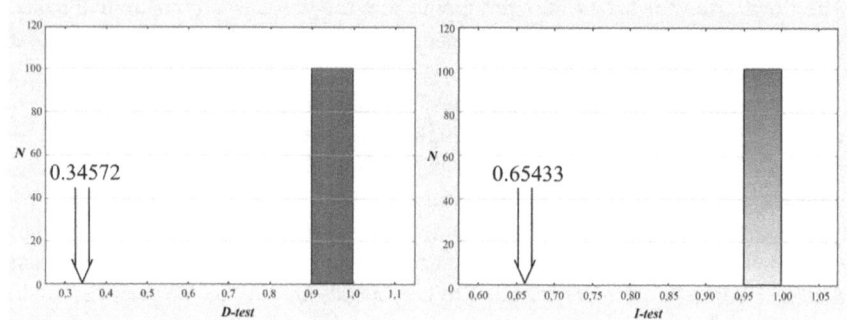

Fig. 5. I-тест–similarity of ecologo-climatic niches of Poa palustris and P. nemoralis in Asian Rusia. **A** – the standardized Hellinger distance. **B** – the Schöner index.

It shows the potential distribution of the species and the gradation of colors, from dark to light, determines the areas where this species can grow, and where it can spread in the future [3].

A darker tone on the map indicates areas with the most favorable combinations for each type of climatic characteristics. Since MaxEnt is currently recognized as one of the best algorithms for modeling the range of species, it is dynamically developing. Developers are taking into account the new aspects, and new opportunities that significantly improve the quality of the model. Such additional features are implemented in the SDMtoolbox application [23, 24] which is a freely distributed application to the ArcGIS licensed program [25]. Improvements relate both to training sampling and climatic characteristics.

In order to improve the quality of the training sampling, this application takes into account the features of the geographic coordinate system, where the area, enclosed between segments of parallels and meridians, varies in latitudinal direction due to convergence of meridians at the poles. The application also takes into account the uniformity of the selection of the material, since the collection is usually the most intensive near the settlements, along rivers, and generally in more inhabited areas, which distorts the picture of the climate preferences of the species. In addition, the application provides processing, thinning the training sample, depending on the uniformity of the geographical distribution of climatic parameters. The special options are supposed for operation of small samplings [26, 27].

All this allows you to more accurately determine the climate profile of the species and improves the quality of the model. Another problem is the correlation between bioclimatic data. As the developers of the program indicate, a high correlation between bioclimatic characters makes the evaluation of their contribution very difficult, and sometimes impossible. The SDMtoolbox application reduces the correlation between climatic characteristics by identifying and removing highly correlated variables.

The program provides an assessment of the predictive capability of the model, namely, to predict the finding of a species where it is, according to climatic needs, occurs, and its absence where it can not grow. This is of great importance in studying the ranges of invasive species and predicting their distribution, as well as in plant introductions. The model is estimated using the Receiver Operating Curve (ROC) and the area under this curve—Area Under the Curve (AUC). ROC—the graph, which allows to estimate the quality of binary classification [12, 26, 28].

It reflects the ratio between the total number of characteristic carriers, correctly classified as carrying characteristics (ordinate axis), and the proportion of objects that do not carry characteristics, that are erroneously classified as bearing the characteristic (the abscissa axis). The quantitative interpretation of ROC is given by the AUC indicator—the area bounded by the ROC and the axis of the fraction of false positive classifications. AUC is an estimate of the ability of the model to indicate the presence of a species at the point of the raster where it is most likely to be located.

AUC is a measure of the area under ROC, ranging from 0.5 (random accuracy) to 1 (ideal discrimination). If the AUC is equal to or below 0.5, the model has no predictive value. A value of 0.8 for AUC means an 80% probability that, where the species is predicted to be, it will indeed be located [10].

The method also makes it possible to evaluate the contribution of each climatic variable in the obtained model of the species distribution, therefore we can evaluate the role of each biologically significant factor, included in the analysis. So, we receive a valuable environmental information that characterizes the species. The evaluation of the contribution of each variable in MaxEnt can be made in three independent ways: direct evaluation of the contribution as a percentage, evaluation after permutation and using the jackknife option. The most important are the last two. When permutation, the contribution of each variable is determined by randomly changing the value of this variable in the analysis, including all the data participating in the "learning" process: both the real presence points of the species, and the "background" ones. In this case, the value of each variable is expressed in a change in the evaluation of the obtained model. The more the estimation decreases, the more the model depends on this variable. In order to express this influence in percent, the values are normalized. The jackknife option consists of three steps. First, all variables are exclude from the analysis in turn, and the model is created with the remaining variables. Then the model is created with only one variable (each in turn). Finally, for comparison, a model is created with the participation of all variables.

Figure 6 presents a model of the potential distribution of *Poa palustris* in the territory of Asian Russia, taking into account the threshold of 10 percentile. The predictive capability of this model is estimated as good—the values of AUC-training and testing were 0.887, and 0.871, respectively. Both characteristics exceed 0.8, which corresponds to a good estimate [10]. The standard deviation was 0.016.

In Fig. 4 a richness of morphotypes of the *Stenopoa* section in the territory of Asian Russia is shown [29]. The widespread introduction of molecular methods into practice led to the development of phylogeographic studies and, as a result, the development of new GIS programs. At present, the applications for mapping of the spatial distribution of genetic diversity [10] and reconstruction of the most probable migration routes [23, 24]. have been developed.

In this brief overview not all the possibilities of using modern GIS technologies in botanical studies are listed, but undoubtedly that their application opens a new page in ecological and geographical studies of species (Fig. 7).

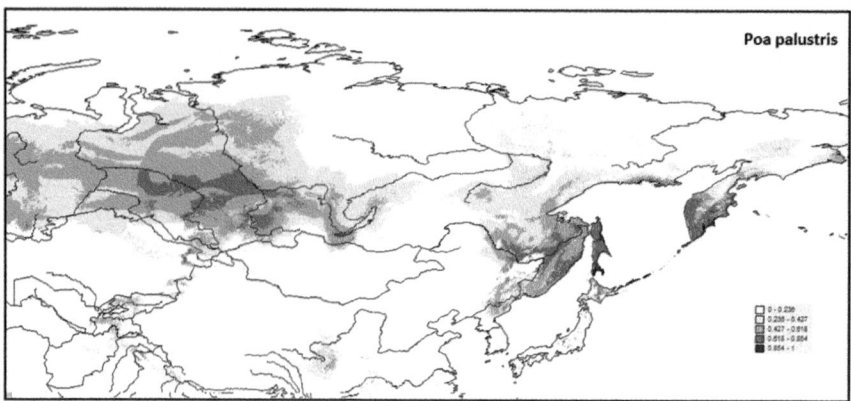

Fig. 6. The model of potential distribution of *Poa palustris* in Asian Russia, obtained in MaxEnt for the current climate, based the ecologo-climatic variables Bio1, Bio2, Bio5, Bio7, Bio8, Bio12, Bio15.

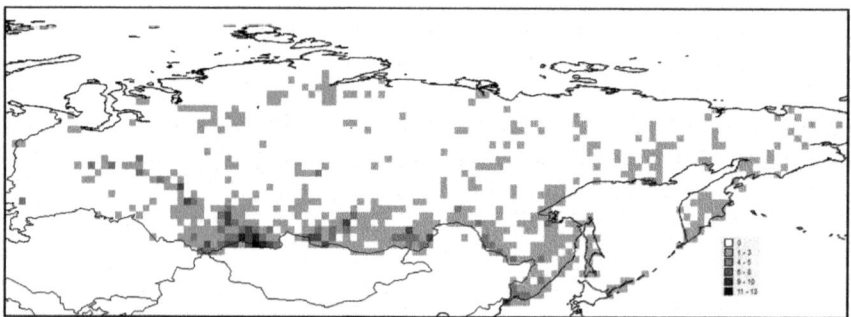

Fig. 7. The richness of flora of Asian Russia with morphotypes of bluegrasses of section Stenopoa.

The research was supported by RFBR, grants № № 15-29-02599 & 16-04-01605 and Mendeleev fund of Tomsk State University.

References

1. Goodwin, B.J., McAllister, A.J., Fahrig, L.: Predicting invasiveness of plants species based on biological information. Concerv. Biol. **13**, 422–426 (1999)
2. Welk, E.: Constrains in range predictions of invasive plant species due to non-equilibrium distribution patterns: Purple loosestrife (*Lythrum salicaria*) in North America. Ecol. Model. **179**, 551–567 (2004)
3. Ward, D.F.: Modelling the potential geographic distribution of invasive ant in New Zeland. Biol. Invasions **9**, 723–735 (2007)
4. Hellmann, J.J., Byers, J.E., Bierwagen, B.G., Dukes, J.S.: Five potential consequences of climate change for invasive species. Conserv. Biol. **22**(3), 534–543 (2008)

5. Hutchinson, G.E.: Concluding remarks. Cold Spring Harb. Symp. Quant. Biol. **22**, 415–422 (1957)
6. MacAthur, R.H.: The theory of the niche. In: Lewontin, R.C. (ed.) Population Biology and Evolution, pp. 159–176. Syracuse University Press, Syracuse (1968)
7. Khlebosolov, E.I.: The theory of niche in ecology: the history and modern status. Russ. Ornithol. J. **11**(203), 1019–1037 (2002)
8. Hijmans, R.J., Guarino, L., Jarvis, A., O'Brien, R., Mathur, P., Bussink, C., Cruz, M., Barrantes, I., Rojas, E.: DIVA-GIS, version 5.2. http://www.diva-gis.org/DIVA-GIS5_manual.pdf. Accessed 18 Apr 2018
9. Franklin, J.: Mapping Species Distribution: Spatial Inference and Prediction. Cambridge University Press, Cambridge (2009)
10. Scheldeman, X., van Zonneveld, M.: Training Manual on Spatial Analysis of Plant Diversity and Distribution. Biodiversity International, Rome (2010)
11. Phillips, S.J.: AT&T. Research. A brief tutorial on MaxEnt. http://www.cs.princeton.edu/~schapire/maxent/tutorial/tutorial.doc. Accessed 21 Jan 2018
12. Elith, J., Graham, C.H., Anderson, R.P., Dudik, M., Ferrier, S., Guisan, A., Hijmans, R.J., Huettmann, F., Leathwick, J., Schapire, R., Soberso, J., Williams, S., Wisz, M., Zimmerman, N.: Novel methods improve prediction of species' distributions from occurrence data. Ecogeography **29**, 129–151 (2006)
13. Ramirez, J., Bueno-Cabrera, A.: Working with climate data and niche modeling. Creation of bioclimatic variables. http://gisweb.ciat.cgiar.org/GCMPage?docs?tutorial_bcvars_creation.pdf. Accessed 21 Nov 2017
14. Beaumont, L.J., Hughes, L., Poulsen, M.: Predicting species distributions: use of climate parameters in BIOCLIM and its impact on predicting of species' current and future distributions. Ecol. Model. **176**, 250–269 (2005)
15. Philips, S.J., Anderson, R.P., Schapire, R.E.: Maximum entropy modeling of species geographic distributions. Ecol. Model. **190**, 231–259 (2006)
16. Phillips, S.J., Dudic, M.: Modeling of species distribution with Maxent: new extentions and a comprehensive evaluation. Ecography **31**, 161–175 (2008)
17. Hijmans, R.J., Guarino, L., Mathur, P.: DIVA-GIS Version 7.5. http://diva-gis.org/docs/DIVA-GIS_manual_7.pdf. Accessed 25 Jan 2018
18. Hijmans, R.J., Cameron, S., Parra, J.: Climate date from Worldclim. http://www.worldclim.org. Accessed 25 Jan 2018
19. Olonova, M.V., Gudkova, P.D.: Bioclimatic Modeling. Tutorial for Students. Tomsk State University, Tomsk (2017)
20. Warren, D.L., Glor, R.E., Turelli, M.: Environmental niche equivalency versus conservatism: quantitative approaches to niche evolution. Evolution **62**, 2868–2883 (2008)
21. Warren, D.L., Glor, R.E., Turelli, M.: ENMTools User Manual v 1.3. http://www.danwarren.net/enmtools/builds/ENMTools_1.4.3.zip. Accessed 13 Apr 2018
22. Zink, R.M.: Genetics, morphology, and ecological niche modeling do not support the subspecies status of the endangered Southwestern Willow Flycatcher (Empidonax traillii extimus). Condor **117**(1), 76–86 (2015)
23. Brown, J.L.: SDMtoolbox User Guide. http://www.jasonleebrown.org/SDMtoolbox/current/User_Guide_SDMtoolbox.pdf. Accessed 15 Mar 2018
24. Brown, J.L.: SDMtoolbox: a python-based GIS toolkit for landscape genetic, biogeographic and species distribution model analyses. Methods Ecol. Evol. **5**(7), 694–700 (2014)
25. ArcGIS (ESRI) Desktop and Spatial Analyst Extension: Release 10.1. Environmental Systems Research Institute, 2012. Redlands, CA. http://www.esri.com. Accessed 16 Apr 2018

26. Anderson, R.P., Lew, D., Peterson, A.T.: Evaluating predictive models of species' distributions: criteria for selecting models. Ecol. Model. **162**, 211–232 (2003)
27. Shcheglovitova, M., Anderson, R.P.: Estimation optimal complexity for ecological niche models: a jackknife approach for species with small sample sizes. Ecol. Model. **269**, 9–17 (2013)
28. Araújo, M.B., Pearson, R.G., Thuiller, W., Erhard, M.: Validation of species-climate impact models under climate change. Glob. Change Biol. **1**, 1504–1513 (2005)
29. Olonova, M.V., Mezina, N.S., Gorina, N.V., Vysokikh, T.S.: Phenetic variability of bluegrasses (Poa L., *Poaceae*) section *Stenopoa* Dum in Asian Russia. Int. J. Environ. Stud. **74**(5), 715–723 (2017)

Diatom Analysis Using SOQL Language Interpreter

Y. V. Avramenko[1($\boxtimes$)], R. K. Fedorov[1], and A. D. Firsova[2]

[1] Matrosov Institute for System Dynamics and Control Theory of the Siberian Branch of the Russian Academy of Sciences, Irkutsk 664033, Russia
avramenko@icc.com, fedorov@icc.ru
[2] Limnological Institute Siberian Branch of the Russian Academy of Sciences, Irkutsk 664033, Russia
firsova@lin.irk.ru

Abstract. Article proposes the method based on logic programming image recognition in order to solve the diatom selection problem. This method is suitable for diatoms selection, as it allows describing every type of diatoms and performing the search. This analysis is based on the study of micro algae organisms, particularly counting the number of diatoms remaining. Counting is usually performed manually by the microscope operator, so, in this paper, we suggest using the SOQL interpreter to find objects to help the operator to estimate the amount of remaining diatom algae.

Keywords: Deformable models · Pattern recognition · Diatoms

1 Introduction

One of the methods of monitoring the ecological state of Lake Baikal is the analysis of diatom algae. Diatom algae belong to the leading group of phytoplankton participates in the food chain and reacts to changes in environmental conditions. The process of estimating the amount of diatom algae in Lake Baikal is performed by the operator and takes a long time. The amount of diatom algae is taken from a sample obtained with a microscope. This process consists of a sequential visual analysis of the magnified parts of the original image. Existing methods and programs implementing them, designed to process data and structural characteristics of phytoplankton, are mainly oriented to a certain set of microalgae. They are equipped with various filters to improve the perception of the image. Here are descriptions of some methods and software.

1.1 Methods

In [1], the Center Supported Segmentation (CSS) method is proposed for segmenting objects on an image by calculating the center of mass of the regions. The method is more accurate than segmentation by a threshold or a watershed method. The limitation of the method is the absence of the possibility of classification.

In [2], a method for detecting a set of overlapping objects using a HOG-SVM detector with modification is presented. The HOG-SVM detector operates on the basis

© Springer Nature Switzerland AG 2019
I. Bychkov and V. Voronin (Eds.): *Information Technologies in the Research of Biodiversity*, SPEES, pp. 169–173, 2019.
https://doi.org/10.1007/978-3-030-11720-7_22

of a sliding window. The modification consists in that the examined image is rotated by a given angle and the method is started again, the process finishes when a lot of classification results are obtained. Further, these results are compared, and for each position of the found object the best option is chosen. The limitation of the method is that it is capable of finding one class of objects at a time and requires training.

The paper [3] gives an overview of the existing methods and the PC discusses their advantages and disadvantages.

1.2 Software

FlowCam. FlowCAM [4] is designed for quantitative calculation of micro-particle volumes. Images are obtained by photographing the flow of liquid containing particles. In the flow, the particles are separated by a sufficient distance, which allows them to be calculated. The program allocates micro-particles on a clean background with a bounding rectangle, determines the dimensions: length, width, diameter. For each particle a descriptor is created. Then a histogram of frequency of occurrence of each type of particles is generated. User can select a range on the histogram and get a list of particles of interest. The program allows partial classification, but does not take into account the shape of the particles, assuming that all particles are round or oval.

Altami Studio. Altami Studio software [5] is designed to manage digital cameras, perform measurements and automatically analyze images. The calibration is required beforehand in order to accurately determine the size of the objects being examined. All operations on the analysis of objects are performed by operators manually.

ImageJ. ImageJ [6] is an open source program for image analysis and processing which allows calculating: area, statistical indicators of pixel values of the area of interest. The selection is made by the user alone or with the help of tools based on threshold functions.

1.3 Summary

The use of existing methods and software for solving this problem is not possible due to a number of limitations. Some of them work only with a certain microscope model. Others are not capable of producing diatom algae. All software is not capable of automatic classification of algae by species. The complexity of this problem lies in the fact that the algae have different shapes (round, rectangular, needle-like and others), size and texture.

2   Proposed Method

2.1   Theory

On the basis of the conclusions formulated in the previous section, it is proposed to use the existing logical method for searching for objects, developed at the ISDCT SB RAS. In order to understand the principles of the proposed method, let us consider the theoretical foundations from the manual on the determination of the biomass of phytoplankton species of the pelagic lake of Lake Baikal. Descriptions of different species are determined by the size of geometric figures (cone, cylinder, parallelepiped, etc.), as well as features of the shell structure. In a simplified form, descriptions are used in the form of geometric figures on a plane (see Fig. 1).

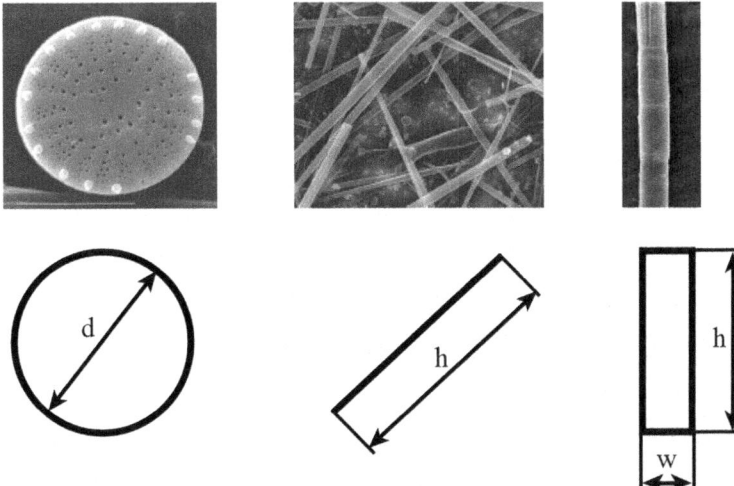

Fig. 1. Types of diatoms and their descriptions.

The method of searching for objects, presented in the current work, allows constructing such descriptions, so it is suggested to use it to help the operator to take into account the amount of diatom algae.

2.2   Implementation

The description of objects is built in the form of a rule in the SOQL language. The form of diatom algae is described by the mutual arrangement of graphical primitives: lines, circles or their combinations. So with the help of combinations it is possible to describe more complex forms, for example a polygon. The length and width are determined by the size of the graphic primitives, and the texture, previously specified by the standard. Example of a rule for searching for diatoms in the form of a rectangle diatom (A, B, C, D):- line(A, B), line(B, C), line(C, D), line(D, A), dist(A, B) > 25, dist(A, B) < 40, dist (B, C) = 60, dist(B, C) = dist(A, D), angle(A, B, C) = 90, angle(B, C, D) = 90, angle (C, D, A) = 90, texture(A, B, C, D, diatom_texture), where: diatom () is the name of the rule; (A, B, C, D) is the set of points in the image; line (X, Y) - a linear segment of the contour between the points X, Y; dist (X, Y) is the distance between the points X, Y; angle (X, Y, Z) is the angle formed by the points (X, Y, Z), texture (W, X, Y, Z, sample) is the sample reference pattern bounded by the set of points W, X, Y, Z. The operation principle of the object search algorithm is as follows. The user sets the threshold value of the energy function (necessary to determine the correspondence of the found object to the description). Then, with the help of a search, features are extracted consistently from the image. As soon as a combination of features satisfying the rule is found, the value of the energy function is calculated for it: if it is less than the threshold value, then the object is considered to be found. This process continues until all possible combinations of attributes that satisfy the rule are considered.

3   Conclusion

The result of applying the proposed method is shown below (see Fig. 2). During the experiments, it was observed that the method allows: in the shortest possible time to create a descriptions of different types of diatoms and to search for them; to process noisy, blurred and other types of images. It is worth noting that more research is required to obtain more accurate search results. Application of the presented method will allow automating the process of counting the mass species of diatoms of Lake Baikal and its tributaries using electron microscopy. This will help most accurately and quickly obtain the results of research conducted in Lake Baikal and its tributaries in the productive years of phytoplankton development.

Fig. 2. The result of applying the proposed method to find objects.

Acknowledgement. The work was carried out with financial support: Russian Foundation for Basic Research (grants №: 18-07-00758-a, 16-07-00411-a, 16-07-00554-a); RAS Presidium Program № 27; Integration program SB RAS, ISC SB RAS. Results are achieved using the Centre of collective usage «Integrated information network of Irkutsk scientific educational complex».

References

1. Charlesa, J.J.: Object segmentation within microscope images of palynofacies. Comput. Geosci. **34**, 688–698 (2008)
2. Liu, Y.: Detecting dense crowds of microbes from microscope images in a global optimization framework. Optik **127**, 76–80 (2016)
3. William, V.: Review: automatic particle detection in electron microscopy. J. Struct. Biol. **133**, 90–101 (2001)
4. FlowCam Fluid Imaging Technologies. https://www.fluidimaging.com. Accessed 01 July 2018
5. Altami Software. http://altamisoft.ru/products/altami_studio. Accessed 01 July 2018
6. ImageJ. https://imagej.net/Welcome. Accessed 01 July 2018

Database of Barguzinsky Reserve

Evgeniya Bukharova[(✉)] [iD], Alexander Ananin [iD],
and Tatiana Ananina [iD]

FSE United Administration of Barguzinsky State Nature Biosphere Reserve and
Zabaikalsky National Park ("Zapovednoe Podlemorye"), Ulan-Ude 670045,
Russia
{darakna, a_ananin, t.l.ananina}@mail.ru

Abstract. One of the objectives of the nature reserves is to a long-term monitoring of natural complexes' state. Barguzinsky Reserve being the oldest in
Russia (organized in 1916) and located on the northeastern coast of Lake Baikal
is a standard of unspoilt nature. Regular observations of natural objects and
phenomena in Barguzinsky Reserve cover the period from the late 1930s.
Currently there is a continuous dataset of observations for 30–60 years. This
data contains: inventory lists of flora and fauna; the results of accounting for the
number of major species of a ground beetles, mammals and birds; an assessment
of the yield of berry and tree species; phenological observations and climatic
data series; observations of populations of rare plant species. The data is stored
in 25 databases. All available materials were used as a source for databases,
including published data, primary observation files, manuscripts of Nature
Annals, and other unpublished reports. Digitization of data makes possible to
systematize the primary observation materials available in the zapovednic, to
facilitate access to them, to preserve these observations from loss, and to perform an inter-related analysis of long-term series of observations. Presently
inventory lists of higher vascular plants and herbarium of the Barguzinsky
Reserve are being prepared for publication in GBIF.

Keywords: Nature reserves · Long-term monitoring · Database

1 Introduction

The increasing scale of the impact of human activities on the environment poses the
problem of the possible consequences of such intervention. The system of ecological
monitoring of the environment created in our country is aimed at solving this problem.

The oldest in Russia Barguzinsky state natural biosphere reserve was organized in
1916. It is locating in the northeastern Baikal region, on the territory belonging to the
background region of Lake Baikal. It is a part of the UNESCO World Heritage Site
since December 1996. It is interesting as a territory—a standard of wild nature in the
Baikal region, never subjected to human exposure.

For a 100-year period of existence in reserve, a considerable scientific material has
been accumulated, including long-term series of observations of various biota components and abiotic factors.

© Springer Nature Switzerland AG 2019
I. Bychkov and V. Voronin (Eds.): *Information Technologies*
in the Research of Biodiversity, SPEES, pp. 174–180, 2019.
https://doi.org/10.1007/978-3-030-11720-7_23

Databases containing information on the results of long-term observations of the parameters of the state of natural complexes of the Barguzinsky reserve were made, depending on the complexity of the material to be included, in the formats Access or Exel. This took into account the possibility of free export of Excel spreadsheets to the files of the database management program (Access) and, conversely, for the purposes of statistical processing of data series, filling out spreadsheets from the database in the Access format.

All available materials of the reserve were used to fill in the databases, including published data, initial survey files, manuscripts of Nature Annals and other unpublished reports. Regular monitoring of the majority of parameters has been started since 1984. Earlier observations often interrupted or changed the methods and places of research, which is primarily due to the employee replacement. This makes it difficult to build long, comparable series of observations and reduces the value of accumulated information.

Filling of electronic databases makes it possible to systematize the primary observation materials available in the reserve, to facilitate access to them, to increase the safety of these observations from losses and provides a real opportunity to perform long-term series of observations.

The databases were prepared by different authors at other times. Therefore, they are numerous and in different formats. Currently, work is underway to bring the data to international standards, note them in one database and for further publication.

2 Characteristics of the Prepared Databases for Long-Term Series

2.1 Weather

The "Weather" database (in Excel format) includes daily information on minimum, maximum and average daily air temperature, minimum soil temperature, precipitation, snow cover height, minimum, maximum and average daily air humidity, average daily atmospheric pressure, average daily soil temperature at depth 5, 10 cm, for the period from 1955 to 1999. The information was obtained at the meteorological station "Barguzinsky Reserve" ("Davshe") of the Irkutsk Hydrometeorological Service located in the Davshe village.

2.2 Water

The "Waters" database (in Access format) contains daily information on the level, temperature and volume of water course in the rivers Davshe and Kudalda, collected by workers of gauging stations of the Trans-Baikal Hydrometeorological Service. Hydrological observation post on the river Davshe was opened in 1978, and on the river Kudalda functioned from 1992 to 1996.

2.3 Soils

The "Soil" database (in Access format) contains every decade information about soil moisture and its temperature at 2 horizons (5, 10 cm) since 1985.

Work on the study of soil moisture and temperature was conducted every ten days on a permanent trial plot (the coast of Lake Baikal, a mixed dark coniferous forest, the type of soil—taiga podbour) in the vicinity of the village of Davsha from the moment of clearing the snow cover and throughout the vegetation period. The temperature of the soil horizons was determined with the help of the Savinov thermometers, and the soil moisture content by the weight method.

2.4 Phenology of Plants

The "Plant Phenology" database (in the Access format) includes the results of phytophenological observations at 5 permanent sites in the vicinity of the village of Davsha (78 plant species, 5 phenophases, 22 phenomena) since 1976.

The beginning of the study of plant phenology in the Barguzinsky Reserve started at the 1930s. Unfortunately, these observations were sketchy and did not have a permanent binding on the terrain. Phytophenological observations have acquired a stationary character since 1965 by the method of Beidemann [1].

2.5 Phenology of Birds

The "Bird Phenology" database (in the Exel format) has been prepared since 1940. Data on spring migrations of 124 species and autumn migrations of 128 bird species (157 species total) have been collected. The time of flight of birds (the first spring meeting, the beginning and the end of the mass migration in the spring and autumn, the first meeting on the autumn migration and the last meeting in the fall) were recorded annually in the spring (March-June) and at the end of summer-in the autumn (July-November). The beginning or the termination of this or that phenomenon in the life of birds was determined by regular registration. Phenological observations were carried out on the coast of the Baikal lake, mainly in the neighborhood of the Dawsha village, Northern and Southern cordons of the reserve [2].

2.6 The Calendar of Nature

Database "Nature calendar " (in Access format) was compiled for the period from 1936. It includes information on 275 phenomena of seasonal nature development of the Barguzinsky Reserve. The boundaries of seasons and sub-seasonals are established by the temperature criteria obtained by processing multi-year dates. When calculating the average dates of occurrence of phenotypes in plants, observations were made on permanent phenological sites. The analysis of zoophenological phenomena also covers only the coastal part of the reserve territory. The terminology used in dividing seasons into sub-seasons is traditional for the Barguzinsky Reserve [3].

2.7   Yields of Berry and Cedar

The "Fruiting" database (in the Access format) contains information on the quantitative evaluation of the yield of 5 species of berry plants over the period since 1971. The productivity of berry fruit is controlled by quantitative method at 39 permanent areas in valleys of 5 rivers, in all high-altitude vegetation belts. The focus is on 5 species: red bilberries, bilberries, blueberries, cranberries and siksha (lat. *Empetrum nigrum*). The "Fruiting" database (in the Access format) includes information of fruits and berries of 42 species of plants and mushroomson in the scores for the period since 1938 on observations throughout the territory of the reserve.

2.8   Winter Route Registration of Animals

Database " Winter route registration" (in Excel format), including information on the occurrence of traces of 15 species belonging to 4 families of 3 orders of mammals, in three high-altitude belts (foothills, mountain forest and highlands) for the period since 1967. Winter route registration of animals is held in the Barguzinsky Reserve on a permanent route running along the belt of the foothill plains (61.5 km), the mountain-forest belt (34 km) and the highlands belt (7 km). The total length of the route is 102.5 km. Registration of animals is carried out according to a standard procedure by two groups [4].

2.9   Spring-Autumn Counts of Mouse-like Rodents

The database "Spring and autumn mouse counts" (in the Access format) includes information on the spring and autumn abundance of 3 species of mouse-like rodents for the period since 1963. The relative registration of mouse was recorded twice a year, in spring and autumn, on 8 permanent counting lines in valleys of the rivers Davshe and Tarkulik. Each line exhibited 100 traps per one.

2.10   Winter Records of Mouse-like Rodents

The database "Winter mouse counts" (in Access format) includes information on the absolute in winter abundance of 4 species of mouse at 10 permanent sites for the period from 1974–75. Absolute registration of mouse was carried out in winter at 10 permanent sites in the valleys of the rivers Tarkulik, Davshe and the interfluve Dawshe-Bolshaya. The works were carried out from the moment of establishing a constant snow cover with a depth of at least 20 cm. In the forest, traps were exhibited on sections of 0.25 hectares.

2.11   Registration of Bears

Database "Bear registration" (in Excel format), including information on the occurrence of a brown bear on the coast of the lake Baikal in the period from 1985. Visual account of the bear, due to the peculiarities of its ecology on the coast of the lake Baikal is carried out from the boat in the early morning and evening hours at 9 stationary points along the protected shore (30 km). The database "Brown Bear", (in format Access), includes a variety of information about all the meetings of the bears in the reserve area since 1957.

2.12   Ungulates

The database "Ungulates" (Access format) includes information on the sex and age structure of 3 ungulate species (elk, reindeer and red deer for the period since 1935. The database uses the materials of the primary observation card files from all reports of the reserve's employees about ungulates' meetings in the protected area.

2.13   Wolf

The "Wolf" database (in the Access format) includes information on the gender, age, and composition of wolf victims since 1935. The database uses materials from the primary observation card files for combines all the information.

2.14   The Monitoring of the Barguzinsky Sable Population

The "Sable" database (in the Access format) contains information on 456 individually tagged sables and 750 catch of live sables. The database includes information about the date and the place of tagging and repeated catching of sables, distance from the place of tagging, sex, age, weight, description of fur color, molting, degree of teeth erosion, condition of genital organs and lactiferous glands of females. The author of the information and the compiler of the DB is Chernikin [6].

2.15   Summer Comprehensive Route-Based Bird Counting

16. Summer complex route recording of birds. The database "Summer counts of ter-restrial birds" (in Excel format) includes information on the number of 114 bird species in 14 plots in the valleys of the rivers Ezovka, Bolshaya and Davshe for the period from 1984. The summer route registration of birds was carried out on permanent routes along the valleys of the rivers Yezovka, Bolshaya and Davshe with a total length of 101 km in the period from 10 June to 5 July. Calculation of population density was carried out according to Ravkin [6].

2.16   Winter Integrated Route Recording of Birds

The database "Winter counts of terrestrial birds" (in Excel format) includes information on the number of 26 bird species in 4 plots in the valley of the river Ezovka since 1984 [6].

2.17   Accounting for Chicken Birds

The database for the autumn recording of chicken birds (in the Access format) includes information on the number of grouses and stone grouse for the period from 1982. Registration of chicken birds were conducted on the 3 regular routes in the valleys of the rivers Ezovka, Bolshaya and Davshe, each year totaling 101 km [7, 8].

2.18   Accounting for Colonial Waterbirds

The database "Registration the common tern" (in Excel format) includes materials on absolute accounting of nesting of the common terns in the territory of 4 colonies since 1984. The counting of nesting common terns was carried out 1–3 times from June 10 to July 10, in 4 colonies by the method of continuous counting of nests, colonies are located on islands and in the estuarine areas of the rivers of the protected coast of Lake Baikal.

2.19   Monitoring of Birds of Prey

The "Birds of Prey" database (in Excel format) contains information on the monthly occurrence of 20 species of daytime birds of prey and 9 species of owls since 1984. A database of primary observations is used in the database, which receives information from all employees of the reserve.

2.20   Accounting for the Number of Beetles

The "Carabidae" database (in Access format) includes information on the occurrence of 99 ground beetles at 17 permanent sites since 1988. The number and species composition of ground beetles (Coleoptera, Carabidae) were studied on 17 permanent entomological lines located in various biotopes on the ecological profile along the valley of the river longer than 30 km, in all high-altitude belts. The method of soil traps was used [9].

2.21   Forest Fires

The database and GIS "Forest Fires" (in ArcGis format) contains information on forest fires for the period since 1952. (place and date of occurrence, area, cause of fire, composition of forest vegetation). The main cause of fires in the territory of the Barguzinsky Reserve is the burning of dead trees when lightning strikes.

2.22   Rare Plants

The database contains data on 14 populations of 10 rare species included in the Red Data Books of different rank. Since 2008, the number, density, age structure, viability of populations is noted [10].

3 Conclusion

In the Barguzin Reserve, databases (in the formats Access and Excel) containing information on long-term dynamics (minimum for 10 years, maximum for 80 years) of controlled parameters of natural complexes have been prepared. Data bases have been prepared in the following sections of long-term observations: weather, water, soil, phenology of plants and birds, nature calendar, yield of berry and tree species, state of populations of rare plant species, winter route registration of animals, spring-autumnal relative the registration of bears, the structure of bear populations, the wolf and ungulate populations, the monitoring of the Barguzin sable population, the summer and winter comprehensive routing surveys of terrestrial birds, the autumn counts of chicken birds, the counting of colonial water birds, the occurrence of birds of prey and owls, ground beetle counts, forest fires. The materials obtained are the basis for assessing the natural variability of the controlled parameters of the state of components of natural complexes and about SIC analysis of long-term observations in order to optimize the monitoring program in the Barguzin biosphere reserve. In addition, this information is used in the development of GIS for protected areas and for assessing biota responses to long-term climate change. Presently inventory lists of higher vascular plants and herbarium of the Barguzinsky Reserve are being prepared for publication in GBIF.

References

1. Beydeman, I.N.: A methodology for studying the phenology of plants and plant communities. Science, Novosibirsk (1974)
2. Filonov, K.P.: Seasonal development of nature in the Barguzin Reserve. Pap. Barguzinsky State Nat. Reserve 7, 47–67 (1978)
3. Filonov, K.P.: Flight of birds in the Barguzinsky Reserve is evidence of seasonal rhythms in nature. Pap. Barguzinsky State Nat. Reserve 5, 30–51 (1967)
4. Teplov, V.P.: Instructions for winter tracing traces. Methods of accounting for the number and geographical distribution of terrestrial vertebrates. Publishing House of the Academy of Sciences of the USSR, Moscow, 342 p. (1952)
5. Chernikin, E.M.: Tagging of Barguzin sables. Bull. MOIP 5(85), 10–23 (1980)
6. Ravkin, Y.S.: To the method of recording birds in forest landscapes. Nature of foci of tick-borne encephalitis in the Altai, pp. 66–75. Science, Novosibirsk (1967)
7. Kiselev, Y.: Features of routing accounting of grouse. Hunting and Hunting. Household 12, 10–25 (1977)
8. Semenov-Tyan-Shansky, O.I.: Organization and methodology of bird counting. Methods Quant. Regist. Game Anim. 5–15 (1964)
9. Grjuntal, S.Y.: Method of quantitative registration of ground beetles. Entomol. Rev. 61(1), 201–205 (1982)
10. Bukharova, E.V.: Study and conservation of rare plant species in the Barguzin Reserve. Publishing house of the BSC SB RAS, Ulan-Ude (2014)

Development of Complex GIS Monitoring
of the Angara River

Andrey S. Gachenko[✉] and Alexei E. Hmelnov

Matrosov Institute for System Dynamics and Control Theory of Siberian Branch
of Russian Academy of Sciences, Irkutsk, Russia
gachenko@icc.ru

Abstract. This work is devoted to development of hybrid geoinformation
system for making forecasts of areas of possible flooding in the downstream
water of Irkutsk hydroelectric station with damages evaluation in conditions of
extreme water content for lake Baikal and for effluents of the Angara river. For
3D model construction of the Angara's bed they used Atlas "Map of the Angara
River in Irkutsk hydroelectric power station up to 142 km on scale of 1: 10000"
General plan of "Waterways of the East Siberian basin", Irkutsk, 1994. Due to
lack of a digital model (map), a paper-based map was digitized using vectorizer
program Easy Trace 7.99. Depth contours, marks and bank lines were also
digitized. To combine the data about land and underwater terrain a specific
software was developed using Delaunay triangulation method. For data layers
their role in the construction of triangulation is specified. The layers contain
information about: land relief, standing water levels, shore lines, underwater
topography. To work with the underwater terrain information about standing
water levels is used. Auxiliary triangulation is made according to the data
whereof further information is revealed regarding heights levels to which shore
contours are attached and in reference to which depths are measured.

Keywords: GIS · Flooding · Lake Baikal · River Angara · Surface

In the course of the project execution on calculation of flooding areas in the downstream of Irkutsk hydraulic power station it was necessary to make a terrain model that
takes into account the shape of the river bottom. If the land relief with different levels of
detail is shown on topographical vector electronic maps of different scales, the information on underwater terrain of the Angara River in soft copies wasn't found. Existing
data of some navigation systems on the depths are fragmented and experts comments
are not precise. The only reliable source is an atlas that is distributed in paper form. In
this regard the need of presented maps digitizing and its combining with the existing
topographic base arose.

Digitizing of paper Atlas of the Angara river maps [1] was performed with the help
of freeware Easy Trace 7.99—Vectorizer of cartographic representations. The following has been digitized: depths curves, marks of depth points, the coastline (Fig. 1).

The Atlas sheets are randomly oriented with the aim of the most close-together
arrangement of the river sections on the pages. For their further use we made rotation
and shift to the map's reference system containing a terrain model. To select the
rotation parameters we compared characteristic points of coastlines contours (capes,

© Springer Nature Switzerland AG 2019
I. Bychkov and V. Voronin (Eds.): *Information Technologies*
in the Research of Biodiversity, SPEES, pp. 181–186, 2019.
https://doi.org/10.1007/978-3-030-11720-7_24

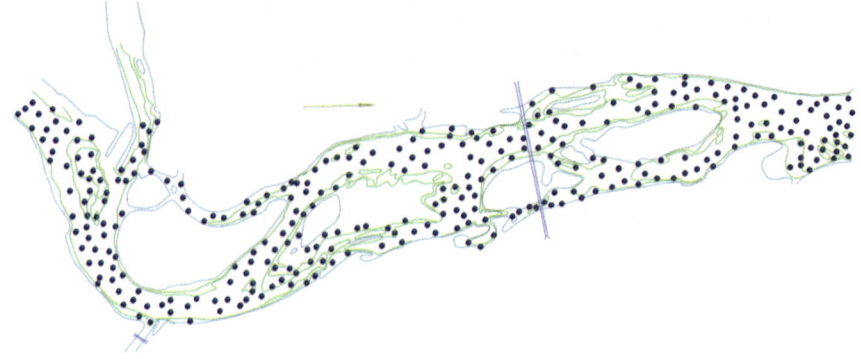

Fig. 1. Digitization result.

bays, islands tips, etc.) of two maps. Utility program was implemented, which calculates the transformation parameters for a given list of points' coordinates pairs and performs map's transformation. To find the transformation parameters they use the least square method. To use the method it is sufficient to specify two pairs of corresponding points, but the use of a larger number of them allows you to get a more exact result, as well as to evaluate the accuracy under the mean square deviation. The mean square deviation of the given examples in this case was 25–70 m. The deviation is due to the presence of significant divergence of contours, because these differences may affect also on characteristic points used for setting of maps conformance.

Attempts to combine contours of vectorized coastline to coastline of topographic base showed that the used atlas was prepared schematically without reference to a map (Fig. 2). That is the atlas reflects the characteristic curves of the shoreline, but it is impossible to combine this line with a more precise contour by turning and shifting.

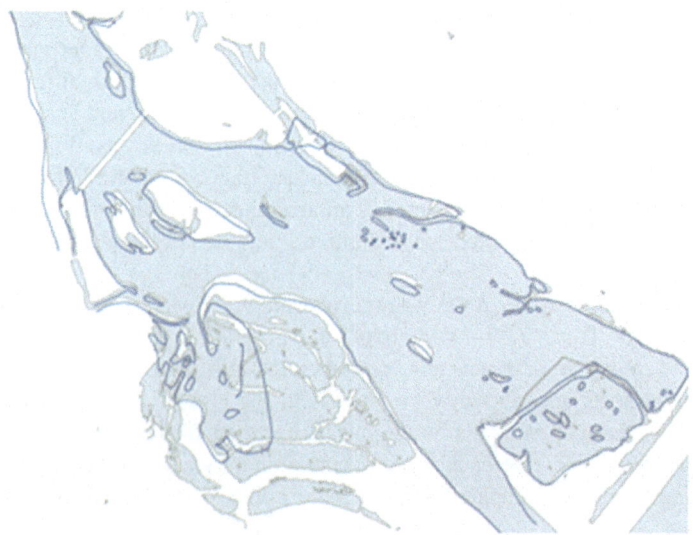

Fig. 2. Maps comparison.

User experience of the calculated under matched characteristics points of rotation and shift showed that the obtained result can not be significantly improved by clarifying the transformation settings.

In order to make possible use of not enough precise data for terrain modeling software for electronic cards morphing have been made. To perform morphing it is necessary to find a continuous transformation of the plate area that can combine inexact contours of coastlines with more accurate. Layers of coastlines are used to align the maps because they exist on both cards. After that, the same transformation is applied to the other layers of the underwater contours maps (isobathic lines and depth marks). The resulting data are better consistent with the information about land relief, at least, transformed isobathic lines and depth marks are not beyond the shoreline. The transformed coastlines contours are not used as there are more exact coastlines available, so their deviation from the target lines is admissible, for example, by adding alignment interdiction. Principal is that the zones of contours deviation do not contain data of used layers of inaccurate map. To perform a map morphing two separate software modules are implemented: for semi-automatic alignment parameter setting and for transformation performing (Fig. 4).

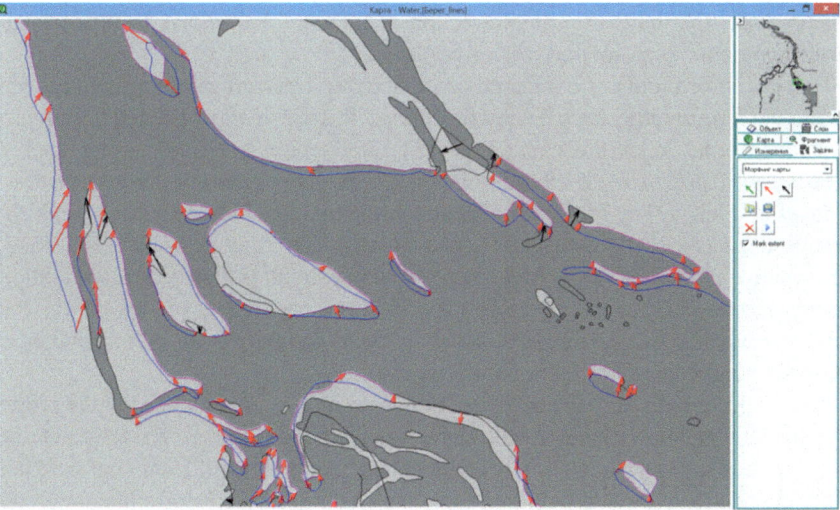

Fig. 3. Contours alignment parameter setting. Accurate map is displayed with filling, movable —as boundary representation.

The contours combination is implemented in the form of a special task in the program of electronic maps viewing IrkGV (Fig. 3). While the task is working the operator sets the directions, combining the maps of starting (inaccurate) and target (exact) maps. When contours combining starting and ending points of drawn by the user arrows are imaged on the closest to them contours points of relevant maps. Further they make a reflection of every contour line of the object of original map to the specified contour area of target map for each pair of arrows, which are adjacent to both

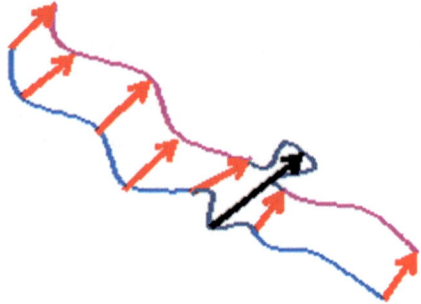

Fig. 4. Maps overlap procedure.

contours. Thus, if, for example, to add an arrow combining this point to the third contour between the points of one contour then the comparison of contours areas stops.

Comparison of two closed loops is performed separately. Without regard to this case it would be necessary to set at least three pairs of corresponding points to connect contours lines totally. While selecting lines for comparison on less number of corresponding points it is needed to be taken into account direction of tracing: they compare contours areas that have the same direction of tracing. This allows to set correlation between contours even by one pair of points.

Displaying of points of compared contour plots is performed using a linear transformation of parametric coordinates of curves. In case if the obtained result is not satisfied, in order to ensure combination of some characteristic contour points it is necessary to match explicitly these points together. After entering the information on the maps alignment they form a displacement file, that in addition to the explicit shifts contains displacements calculated for the intermediate contour points. Blockage of overlapping circuit plots is required when they compare plots represented with different refining degree. For example, there is a specified precise contour of bay, that is absent in the less precise. In this case it is better to avoid comparison of corresponding bay plot with straight line of transformable contour.

To transform morphing the corresponding operation is implemented in the program of triangulations construction. At the same time according to the original points from file of displacements triangulation [2] with constraints is formed. Hard edges are being added for the nearly located points of the same contour. Shift vector to the target point is stored additionally at every point of triangulation. After that, layers processing is performed in Shape format: each point is moved to the vector obtained by linear interpolation from points of triangulation.

For matching data on above-water and underwater terrain they modified form of triangulation construction by map's layers. Now the role at construction of the triangulation is indicated. A layer can contain the following data: land relief, water edge mark, contours of coastlines, underwater terrain. To work with underwater terrain it is required to provide information about water edge marks. According to the data we construct auxiliary triangulation from which further information is extracted about altitude marks, to which coastline contours are attached and with regard to which depths are measured.

To calculate flooded areas they use contour lines construction according to triangulation, that points altitude is determined by the difference in altitudes of points on the original terrain and calculated with level of water in case of flooding. This approach allows to get more accurate results and to cover larger area than construction of contour lines according to terrain model for the altitude equal to the level of water at any point increased for the height of water rise at this point that is often used in such cases (Figs. 5 and 6).

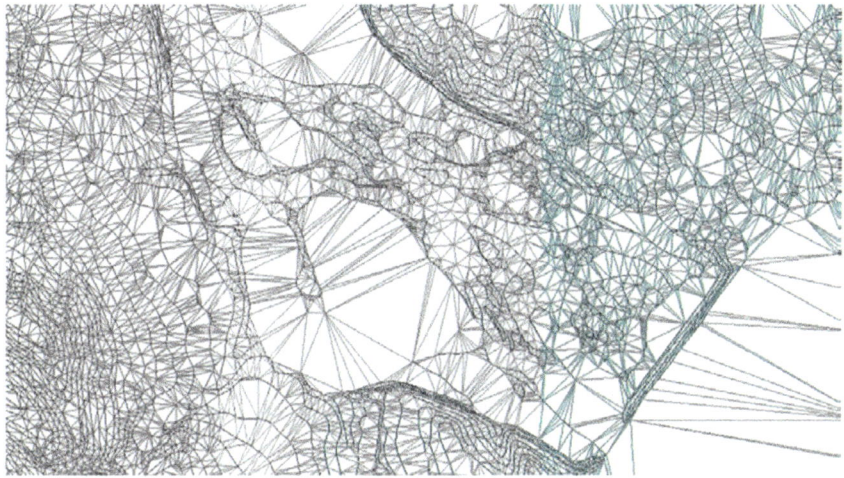

Fig. 5. Segment of formed triangulation, containing information on land and underwater terrain.

Fig. 6. Combined terrain (altitude is increased for clarity).

Formed terrain model is stored in a triangulation file (files extension .trg) and can be used to perform further analysis using dynamic library TrigLib.dll, designed to work with this data. This library allows to obtain the height of the terrain from triangulation at any point and to perform line tracing, i.e. to receive terrain section along a segment or a broken line.

In addition to the basic triangulation, containing information about the terrain, auxiliary triangulation built according to water edge marks is also formed, which contains information about the water level surface used for the construction of a generalized terrain model.

The program of construction of triangulation includes morphoprofile generation algorithms under specified in the file coordinates, terrain contouring under the given triangulation, and subtraction of altitudes given by one triangulation from the heights of the other points of triangulation. Thus, for the construction of flood zones they make triangulation that takes into account the calculated water level.

On the basis of use of dynamic library TrigLib.dll a program for altitude marks forming according to randomly given sort of section of the Angara was developed.

Conclusion. As a result of performed works on evaluation of extreme floods under different scenarios of extreme water availability in the basin of the Angara river and lake Baikal, they created hybrid geoinformation system [3–5], which allows to simulate different scenarios of floods and to identify flooded area considering underwater and land relief.

References

1. Map of the Angara river from Hydroelectric power station of Irkutsk till 142 km, The Ministry of transport of the Russian Federation, Department of inland transport, Glavvodput (Main water way), State Enterprise «Water routs of East Siberian Basin», Irkutsk (1994)
2. Skvortsov, A.V.: Delaunay Triangulation and Its Appliance. Tomsk University press, Tomsk (2002)
3. Fereferov, E.S., Bychkov, I.V., Gachenko, A.S., Popova, A.K., Rugnikov, G.M., Hmelnov, A.E.: Employment of GIS- and Web-technologies for integrated information analysis systems. Comput. Technol. **12**(3), 5–18 (2007)
4. Bychkov, I.V., Kuznetsova, T.I., Batuev, A.R., Pliusnin, V.M., Rugnikov, G.M., Hmelnov, A.E.: Structured-typological characteristics and ecological potential of geosystems of Baikal region. Geogr. Nat. Resour. **4**, 20–28 (2011)
5. Bychkov, I.V., Plyusnin, V.M., Rugnikov, G.M., Fedorov, R.K., Hmelnov, A.E., Gachenko, A.S.: The creation of a spatial data infrastructure in Management of regions (exemplified by Irkutsk oblast). Geogr. Nat. Resour. **34**(2), 191–195 (2013)

Current State and Rational Use of Landscapes in the Border Area of Mongolia and Russia

Alexey I. Shekhovtsov$^{(\boxtimes)}$, Irina A. Belozertseva ⓘ,
Igor N. Vladimirov ⓘ, and Darya N. Lopatina

V.B, Sochava Institute of Geography SB RAS, St. Ulan-Batorskaya 1, 664033
Irkutsk, Russia
ashekhov@irigs.irk.ru

Abstract. As a result of integrated assessment of natural and anthropogenous situation in Hentyi, Dornod and Darhan-Uul provinces (aimags) we have found, that the forest shrinkage and fires, overgrazing and resulting soil erosion are the most acute environmental problems here. As a result of chemical analysis we have found, that the Onon river waters and its left inflows belong to hydro-carbonate calcic type of the first group there is a low water mineralization of the left inflows and the increased mineralization of the right inflows. The rivers Shusyn-Gol, Arangatyl-Gol and Bayan-Gol are allocated with the increased contents of strontium from all rivers tested on heavy metals. Comparing background contents of soil chemical elements in the area of Russia and Mongolia it is possible to draw a conclusion, that soils in the area of northeast Mongolia and Transbaikalia are comparable, they have a low contents of macro- and microelements, except manganese, strontium and vanadium, which have concentration, exceeding the lithosphere clark in Mongolia. There are elevated contents of lead and cadmium in urban soils on the territory of districts (somons), exceeding maximum concentration limit by 1, 5–2 times; titanium and barium, exceeding the lithosphere clark, that is connected with oven heating and transport. We made an ecological zoning of the area on categories of the importance and sensitivity of soils for their potential development.

Keywords: Landscapes · Soil · Mongolia

1 Introduction

Given that the northern Mongolia by natural conditions is a continuation of the East Siberian landscapes, its territory is subject to the influence of the same catastrophic natural phenomena. A significant role in changing the landscape and biota is also played by human activities and, directly or indirectly related phenomena (fires, soil degradation, etc.) [1]. Harsh environment of Mongolia historically determined the population's inability to engage in productive agriculture and settled cattle breeding. As a result, grazing of cattle, covering large areas, has become the main factor contributing to the development of land degradation and desertification. The mining industry, deforestation and desertification also have a destructive impact on the environment [2].

© Springer Nature Switzerland AG 2019
I. Bychkov and V. Voronin (Eds.): *Information Technologies in the Research of Biodiversity,* SPEES, pp. 187–192, 2019.
https://doi.org/10.1007/978-3-030-11720-7_25

Since 2009 at the insistence of NGOs a special law "On the prohibition of exploration and mining in river sources, water protection zones and forest lands" has been adopted in Mongolia; licenses have been revoked and all the mines and sections that fell into the new security zones have been closed. As a result on the Mongolian part of the Onon river basin there are fewer than a dozen mining companies and most of them can not start work, waiting for the final implementation of the new law.

2 Methods and Object

Landscape-geographical studies were carried out in the territory of northern Mongolia within three aimags (provinces)—Khentyi, Dornod and Darkhan-Uul. Totally more than 100 soil sections have been laid, and about 400 samples of water and soils have been selected. Chemical analyzes of soils were determined by generally accepted methods in the licensed chemical analytical center of the V.B. Sochava Institute of Geography SB RAS (IG SB RAS). Water and soil analyzes for the content of gross form of macro- and microelements were carried out by quantitative spectrometric methods on DFS-8 and atomic-emission Optima 2000DV devices.

3 Results of a Research

There were landscape and geographical research works in the area of northern Mongolia within three aimags (Khentyi, Dornod and Darhan-Uul). As a result of pre-integrated assessment of natural and anthropogenous situation in Hentyi, Dornod and Darkhan-Uul aimags we have found, that the forest shrinkage and fires, overgrazing and resulting soil erosion are the most acute environmental problems here [3]. For the last 40 years the forest area in Mongolia reduced by 1.4 million hectares, and 42.5 million hectares of the area more or less prone to desertification. Also we determined various steps of vegetation pasture degression on a much larger area.

As a result of chemical analysis we have found, that the Onon river waters and its left inflows belong to hydrocarbonate calcic type of the first group, with a low mineralization (an ions ratio are defined as $HCO^{3-} > Cl^- > SO_4^{2-}, Ca^{2+} > Na^+ + K^+ > Mg^{2+}$). The water of the rivers flowing through the steppe plains, such as Shusyn-Gol, Bayan-Gol, Hurhyn-Gol and Tuul has a high mineral content, reaching up from 270 to 971 mg/l, respectively. So, there is a low water mineralization of the left inflows and the increased mineralization of the right inflows. The rivers Shusyn-Gol (1021 mkg/dm^3), Arangatyl-Gol (885 mkg/dm^3) and Bayan-Gol (537 mkg/dm^3) are allocated with the increased contents of strontium from all rivers tested on heavy metals.

The soil chemical analysis revealed that contents of macro- and microelements in agricultural land soils of northern Mongolia have values close to background and do not exceed maximum concentration limits. However, the background content of manganese and strontium in soils of Mongolia exceeds the lithosphere clark twice, that is connected with soil- forming rocks. There are elevated contents of lead and cadmium in urban soils on the territory of somons (districts) and cities of Ulaanbaatar, exceeding maximum concentration limit by 1, 5–2 times; titanium and barium, exceeding the lithosphere clark, that is connected with oven heating and transport (see Fig. 1, Table 1). Comparing background contents of soil chemical elements in the area of Russia [3] and Mongolia it is possible to draw a conclusion, that soils in the area of northeast Mongolia and Transbaikalia are comparable, they have a low contents of macro- and microelements, except manganese, strontium and vanadium, which have concentration, exceeding the lithosphere clark (percent abundance of elements) in Mongolia. Contents of chemical elements in sediments and in soils are close.

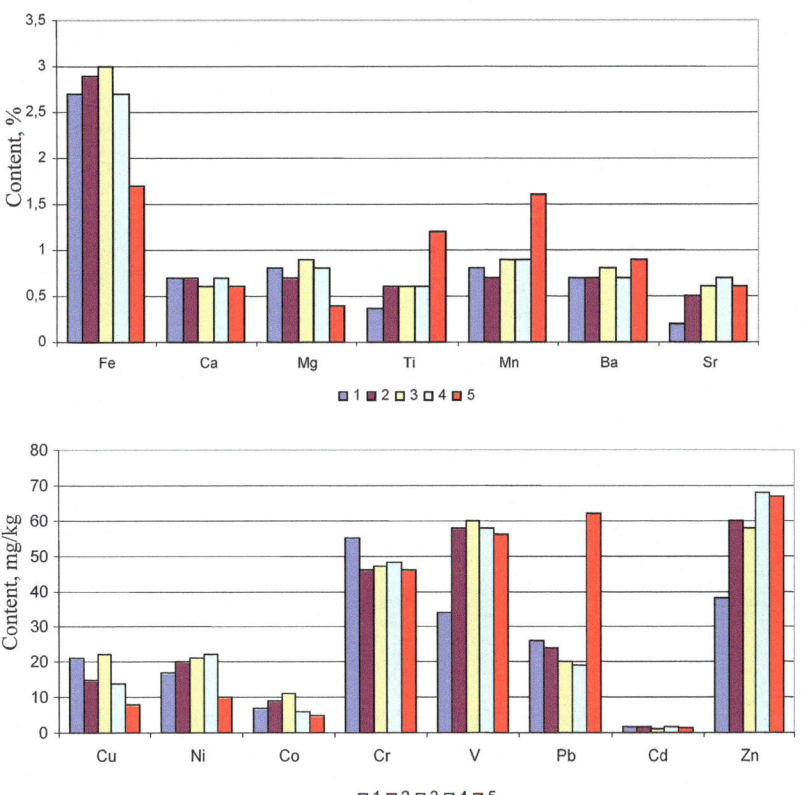

Fig. 1. The average content of chemical elements: background soils of the Onon river basin on the territory of Russia (1), Mongolia (2); sediments (3); soils of arable grounds (4); soils of somons and Ulaanbaatar (5).

Table 1. The content of chemical elements in the soils of Northern Mongolia.

Location	Ti	Mn	Ba	Cu	Ni	Sr	Co	Cr	V	Pb	Cd	Zn
	%	mg/kg										
Russia - Mongolia, near the border	0.2	590	529	8	20	466	4	26	32	25	1.6	70
Background area, 30 km from the border	0.6	692	655	15	20	690	9	46	76	24	1.7	72
Arable land, 42 km from the border	0.6	895	732	14	22	827	6	48	85	10	1.9	68
The right bank of the Onon river	0.3	505	564	11	19	533	3	34	50	23	1.3	74
Somon Dadal	0.2	866	462	6	12	572	6	29	31	22	1.2	82
Floodplain of Balj-Gol	0.4	677	502	14	22	605	7	50	54	20	1.8	56
2 km from the site № 21 to the south	0.3	761	494	15	**45**	559	8	39	49	20	1.6	74
10 km from the site № 21 to the south-west	0.3	435	407	24	24	474	3	44	51	11	**3.7**	68
Valley of Shusyn-Gol, pasture	0.2	340	708	7	25	348	3	36	35	23	1.4	59
Somon Bayan	0.2	403	619	6	19	426	7	19	23	21	1.9	54
Valley of Onon river, pasture	0.4	442	627	11	42	500	3	38	51	26	1.0	85
Valley of Bayan-Gol, pasture	0.3	724	560	11	14	396	15	29	50	29	1.0	57
Valley of Burun-Gol, pasture	**1.1**	681	716	31	25	832	13	69	129	10	3.2	72
Valley of Burun-Gol	0.5	479	541	15	23	756	5	52	100	11	1.2	45
Valley of Barh-Gol	0.2	641	493	9	29	477	5	35	40	21	1.8	65
The southern slope to the valley of Egyin-Gol	0.3	452	445	9	16	435	3	29	47	18	1.5	51
Somon Batshiret	0.3	**1588**	567	8	10	386	5	26	46	**60**	1.3	67
Valley of Amgalant-river, near the village	**1.2**	580	**931**	10	10	446	4	26	44	18	2.2	80
Floodplain of Onon river	0.3	581	549	7	15	383	6	23	42	18	2.1	73
Valley of Bayan-Gol and its inflow Hurh-Gol, pasture	0.4	666	619	16	22	527	8	41	59	18	1.5	77
Valley of Jargalantai-Gol, near the village	0.5	**1009**	350	24	17	296	13	51	71	**62**	1.0	70
The southern slope to the valley of Jargalantai-Gol	0.9	890	874	29	35	520	12	62	93	41	2.0	76
City of Ulaanbaatar	0.3	**1090**	570	10	13	420	5	28	50	**62**	1.5	78

According to the methods of IG SB RAS [4–6] we made an ecological zoning of the area on categories of the importance and sensitivity of soils for their potential development with allocation of zones, such as a zone of pre-emptive conservation of the current state and use of soils, a zone of preservations of the existing sustainable extensive use of soils or transfer to this category, a zone of the primary development of the existing and planned use of soils (see Fig. 2, Table 2).

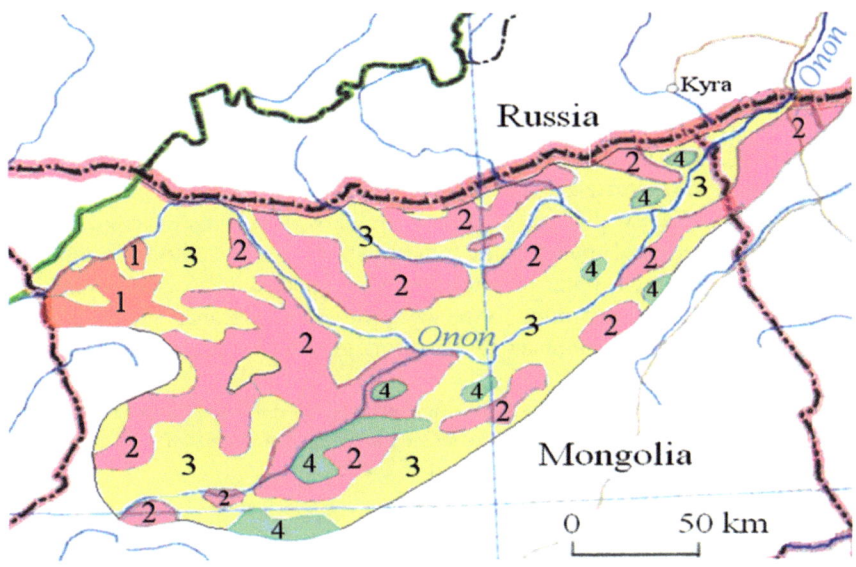

Fig. 2. Ecological zoning of a soil cover of the Onon river basin in Mongolia.

Table 2. The legend of Map «Ecological zoning of a soil cover of the Onon river basin in Mongolia».

	Types of goals		Soils
	Mainly preservation of a current state / use	Refusal of use	Sensitive and high-significant: cryozems, sod-podburs (ashed), peat-podburs, peat-lithozems
		Preservation of the existing sustainable extensive use or transfer to this category	Valuable, possessing good prerequisites on a bioproductivity: dispersed-carbonate chernozems (hydrometamorphic), chernozemlike dark-humic typical (metamorphic). Soils, very sensitive to erosive processes: lithozems, petrozems, podburs. Long-seasonally frozen ground and permafrost: peat-eutrophic (gley), humic-hydrometamorphic, mould-hydrometamorphic, peat-cryozems
	Mainly development of the existing and planned use	Extensive development	Aligned surfaces soils of flat or gently rolling slopes: dispersed carbonate thin detrital chernozems, hydrometamorphic chernozems, dark-humic residual carbonate. Soils sections of the river valleys sections: alluvial gray-humic and dark-humic (gley), stratified, mould-gley
	Mainly improvement / sanitation	Improvement with the subsequent transfers to category of extensive use	The broken soils as a result of intensive environmental management

4   Conclusions

As a result of the research carried out within the boundaries of Khentyi, Dornod and Darkhan-Uul aimags, we have established that forest drying and fires, cattle grazing and associated soil erosion are the most acute environmental problems. On the territory of somons, there are increased concentrations of Pb and Cd in soils exceeding the MAC and UDC; Ti and Ba, exceeding the clark of the lithosphere, which is associated with furnace heating and road transport. In some rivers of northern Mongolia, increased mineralization and a high content of strontium were detected. Ecological zoning of the territory was carried out according to the categories of significance and sensitivity of soils for their potential development. Most of the territory is suitable for land use.

Acknowledgements. The reported study was funded by RFBR according to the research project №. 17-29-05089, by V.B. Sochava Institute of Geography SB RAS research program according to the projects №. 0347-2016-0001 and 0347-2016-0003.

References

1. Vyrkin, V.B., Aleshin, A.G., Belozertseva, I.A., Vyrkina, L.A., Mironova, E.N.: Landscapes of the Darkhat basin (Northern Mongolia). Geogr. Nat. Resour. **2**, 140–148 (2004). (in Russian)
2. Belozertseva, I.A., Dorygotov, D., Sorokovoy, A.A.: Ecological zoning of soils of the Lake Baikal Basin in territory of Russia and Mongolia. Sylwan **159**(8), 319–332 (2015)
3. Vyrkin, V.B., Plyusnin, V.M., Belozertseva, I.A., Shekhovtsov, A.I., Enushchenko, I.V., Zakharov, V.V.: The current status of nature, and ecological problems of the Middle Onon region. Geogr. Nat. Resour. **35**(1), 18–26 (2014)
4. Antipov, A.N., Semenov, Yu, M.: The Russian school of landscape planning. In: Vogtmann, H., Dobretsov, N. (eds.) Environmental Security and Sustainable Land Use—With Special Reference to Central Asia. NATO Security through Science Series, pp. 309–319. Springer, Dordrecht (2006)
5. Belov, A.V., Vladimirov, I.N., Sokolova, L.P.: Cartographic assessment of the present status of vegetation in Prebaikalia for water use optimization. Geogr. Nat. Resour. **37**(2), 129–134 (2016)
6. Plyusnin, V.M., Vladimirov, I.N.: Territorial planning of the Central Ecological Zone of the Baikal Natural Territory. Academic Publishing House "GEO", Novosibirsk (2013). (in Russian)

Environmental Aspects of Urbanized Territories in the Baikal Region

Olga V. Gagarinova[1], Andrey Sorokovoy[1(✉)],
Irina A. Belozertseva[1,2], Natalia V. Emelyanova[1],
and Roman Fedorov[3]

[1] V.B. Sochava Institute of Geography SB RAS, Irkutsk, Russia
geomer@irigs.irk.ru, belozia@mail.ru
[2] Irkutsk State University, Irkutsk, Russia
[3] Matrosov Institute for System Dynamics and Control Theory SB RAS, Irkutsk, Russia

Abstract. Urbanization is a multifaceted phenomenon from the social point of view. On the one hand, the city as a special form of socio-geographical and socio-cultural space organization is a factor in the progressive development of society. On the other hand, the urban processes have a negative impact on the quality of urban environment. As the rate of urbanization increases, the quality of life in the largest cities depends on the degree of natural environment protection. On the basis of statistical data, we consider the main factors of urban processes in the Baikal region. We distinguished the role of geodemographic factors in the transformation of urbanization and investigated the quality of life of the urban population in the Baikal region. The demographic and ecological aspects of the quality of life are illustrated by the example of regional centers (Irkutsk, Ulan-Ude, and Chita).

Keywords: Ecology · Human health · Urbanization · Anthropogenic impact · Water resources · Water quality · Soil degradation · Baikal region

1 Introduction

The specific character of the country is quite well reflected in the geography of Russian cities, i.e. the large size of space, a consistent approach to the formation of the State territory, and the course of urbanization. The regions of the RF differ from one another by the degree of development, the level of socio-economic advancement and matureness of the systems of cities, and the spatial pattern of population distribution [1].

We consider the main factors driving the urban processes in the Baikal region using statistical data, identify the role of the geodemographic factors in the transformation of urbanization and investigate the issues concerning the quality of life in the cities of the Baikal region.

One of the main environmental aspects of the living standard is the water resource quality. An assessment of the level of anthropogenic impact on water resources in the Baikal region has shown that the highest degree of this impact on the hydrosphere is

© Springer Nature Switzerland AG 2019
I. Bychkov and V. Voronin (Eds.): *Information Technologies in the Research of Biodiversity*, SPEES, pp. 193–199, 2019.
https://doi.org/10.1007/978-3-030-11720-7_26

observed in Irkutsk oblast. The main loads are associated with the volumes of water intake and sanitation in cities and industrial centers. We carried out the mapping of the soil degradation and soil contamination in the Baikal region. The territory of the region under investigation includes the Central Siberian, Baikal-Dzhugdzhur and Southern Siberian natural regions. Smaller-scale subdivisions of the territory into the landscape and geochemical provinces are identified by the set of factors of potential soil contamination and degradation hazards posed by nature management. The map shows soil contamination zones with an exceedance of maximum permissible concentration (MPC) of pollutants, their gross emissions, industrial sources and their contribution to atmospheric pollution.

2 Study Area

The Baikal region includes three subjects of the Russian Federation united by their belonging to the Lake Baikal drainage basin: Irkutsk oblast, the Republic of Buryatia, and Zabaikal'skii krai. The total area of the region is 1558.1 thou km^2, and the population size is 4,480,900 (as of January 1, 2016).

The Baikal region is dominated by the urban population living in cities of different socio-economic types. According to the level of urbanization in the region, Irkutsk oblast occupies the first place, followed by Zabaikal'skii krai and the Republic of Buryatia (Table 1). The excess of the urban population in Irkutsk oblast is primarily due to the fact that its industry is much highly developed than agriculture, and its separate branches, such as the wood-processing, aluminum and energy industries, hold the lead in the region. Therefore, the population of Irkutsk oblast is concentrated in industrially developed centers, i.e. in cities.

Table 1. Population and structure of the Baikal region.

	Population, thou, January 1, 2016	7Urban	Rural	Districts	Number of cities	Number of urban settlements	Percentage of urban population, %
Irkutsk oblast	2412.8	1905. 2	507.6	33	22	51	78.9
Republic of Buryatia	982.3	579.4	402.9	21	6	12	58.9
Zabaikal'skiikrai	1083.0	733.4	349.6	31	10	41	67.7
Baikal region	4478.1	3218	1260.1	85	38	104	71.8

Most cities of the Baikal region are small in terms of the population size; Usolye-Sibirskoye, Ust'-Ilimsk, Cheremkhovo, and Krasnokamensk are categorized as medium sized; the cities of Ulan-Ude, Chita, Bratsk and Angarsk are large, and Irkutsk is the largest [2].

The socio-economic situation in the period under consideration (1990–2016) in the cities used in this study was characterized by abrupt changes in all spheres of public life. At the end of the 20th century there was a significant decline and layoff of production, a sharp differentiation in the incomes of citizens, their massive impoverishment, rising unemployment, and economic, legal and moral insecurity worsened. However, an improvement in the socio-economic situation has been observed since the beginning of the 21st century.

The population of the cities is multi-ethnic. In 2010, about 100 ethnic groups of the permanent population were registered in each of them. The majority of Russians, 88%, live in Irkutsk, 62.1% in Ulan-Ude, and 92% in Chita; the share of Buryats (occupying the second place) is 3.2, 31.9%, respectively.

3 Methods

The research technique is based on a comprehensive approach that includes the comparative geographical, statistical, ecological-biochemical and cartographic methods. The information-empirical and normative basis for this research was provided by official methodological and statistical material of Rosstat (Federal State Statistics Service), the Administration of the city of Irkutsk, and thematic statistical bulletins.

4 Results and Discussion

It is now established that the average life expectancy and the other indicators of human health largely depend on the degree of environmental pollution, and the indicators of the health status of the population can be a function of the state of natural environment. Therefore, it is necessary to monitor the current environmental situation, including the assessment and comparison of the quality parameters of individual components of the natural environment with existing standards and conduct special medical and environmental research aimed at identifying environmental factors that adversely affect the health of population, taking into account their integral impact and the different degrees of stress in the environmental situation [3].

The climate in these cities is similar. It is extremely continental and uncomfortable. Geomorphological and atmospheric conditions contribute to the accumulation of air pollutants. This determines the high level of contamination of the ground layer of the atmosphere by more than 200 substances. Thus, 214 harmful substances enter the atmospheric air of Ulan-Ude. A significant air emitter is the road transport and heat power facilities. The highest levels of pollution are observed in the autumn-winter period, which contributes to seasonal heating. The cities of Irkutsk, Ulan-Ude and Chita are constantly included in the Priority List of cities (annually about 40 cities of the Russian Federation) with a very high level of air pollution. Gross emissions, industrial emitters and their contribution to atmospheric emission are presented.

One of the main environmental aspects of living standards is the quality of water resources [4] (Fig. 1).

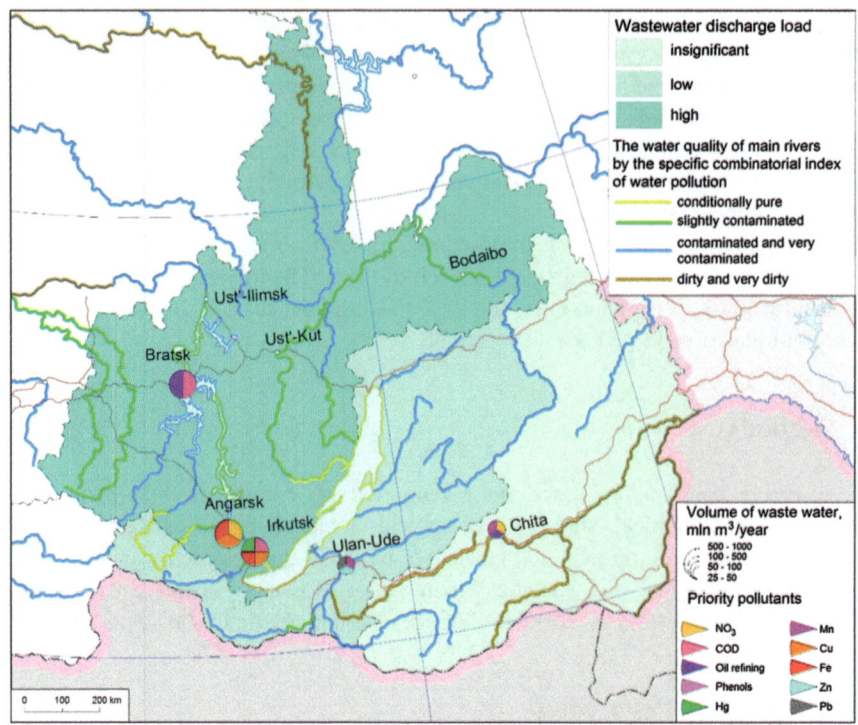

Fig. 1. Impact intensity on natural waters as a result of wastewater disposal.

Comparison of generalized indicators of loads on the surface waters of the Baikal region as a result of socio-economic development shows the following features:

- The volume of water consumed and wastewater discharged is the largest in Irkutsk oblast.
- The main consumers are urbanized territories: the cities of Irkutsk, Angarsk and Bratsk. On the other hand, the water quality in the rivers and reservoirs of Irkutsk oblast is satisfactory, which indicates a sufficiently good degree of wastewater treatment.
- Minimal volumes of wastewater discharged are recorded in Zabaikal'skii krai. However, the poor water quality in rivers is observed, which is due to the activity of mining enterprises and the anthropogenic impact of large industrial centers. In the water of the Chita, Ingoda and other rivers, significant exceedances of the standards for nitrogen content levels, surface active substances, petroleum products, etc. are recorded.
- In the Republic of Buryatia, sewage volumes are relatively high. In the Selenga and Uda rivers in Ulan-Ude, the water quality is low, and exceedances of concentrations of nitrates, iron, fluorine and other substances are recorded.

An important issue in the development of the territory is to provide the population with safe drinking water. The structure of water supply in different parts of the Baikal region is different. In Irkutsk oblast, the water supply is provided exclusively by surface waters, which is due to their considerable volume. On the other hand, drinking water is characterized by very low levels of mineral components. This causes an increase in cardiovascular diseases, disorders of digestive organs, kidneys, the osteoarticular system, and other diseases.

For Buryatia and Zabaikal'skii krai, the opposite situation is typical where drinking water is supplied mainly from underground sources, and water abstraction for industrial purposes is provided by various types of water resources.

Groundwater sources for drinking purposes do not always correspond to sanitary-hygienic standards. In the city of Chita, drinking groundwater has a high level of mineralization, a high level of hardness, and high content levels of iron, manganese and fluoride.

The general analysis shows that the share of the population provided with safe drinking water is larger in the city of Irkutsk and in Irkutsk oblast, next comes Zabaikal'skii krai and then Buryatia [5–7].

All regions are experiencing an intense anthropogenic impact on natural waters, which leads to a deterioration in the drinking water quality. Underground and surface sources of drinking water supplied to large industrial centers often do not meet the hygienic requirements. The largest number of water supply sources that do not meet the hygiene standards is recorded in Irkutsk oblast. The predominant surface water supply can be considered to be the main factor responsible for this situation, because of the vulnerable character of groundwater sources as they are readily exposed to pollution and depletion.

A number of drinking water samples in the centralized water supply system that does not meet hygienic standards shows that the least favorable conditions are observed in Zabaikal'skii krai and, especially, in Chita. The reason for this is largely that water supply networks are in an emergency condition, the deterioration of which makes up 90% in Chita.

One of the factors for the tense ecological situation in the Baikal region is the soil cover. Persistent areas of technogenic pollution are produced in the regional and industrial centers (Irkutsk, Angarsk, Bratsk, Ulan-Ude, Severobaikalsk, Chita and Krasnokamensk). The most significant (in terms of area and intensity of the disturbance of soils and the geological environment) were identified in the Irkutsk coal-bearing basin, the Angara-Ilim iron ore basin, the Gusinoozersk coal deposit, the tailing dumps of the Dzhida tungsten-molybdenum plant, and in the Krasnochikoisk and the Zaka-mensk gold fields [8].

There are areas of soil cover contamination by pollutants exceeding MPC by a factor 1 to 10 as regards the amount of priority toxic chemical elements.

Environmental contamination is the main cause for anthropogenic diseases. For large cities the main sources of the leading pathology forms are waste products (entering in different aggregate states the ground air layer and reservoirs which are often used as water consumption sources) as well as the transport, housing and communal services.

In the 1990s, Irkutsk, Ulan-Ude and Chita experienced a decline in the population. There was a decrease in the birthrate and an increase in mortality, and a decrease in life expectancy, which is an integral indicator of human health. The indicators of natural increase had negative values. This was due to negative socio-economic factors, the layoff of many industries, and to unemployment. The existing differences between cities are related to the level of deterioration of the social and economic situation as well as to differences in the ethnic population structure.

Since the beginning of the 21st century, medical and demographic processes have changed their focus. The birthrate has increased, mortality decreased, and the life expectancy began to increase. However, while the level of fertility, according to the WHO classification, is estimated as low (its value does not exceed 14.9 births per 1000 people), and the mortality rate as average (the average interval is 9–14.9).

The life expectancy since 2007 has been higher in cities than in rural areas. This is determined by the positive factors of cities: better housing conditions, the implementation of programs of improvement of the medical and demographic situation and of preventive measures, and early detection and treatment of diseases that are the main causes of disability and mortality. In the structure of mortality, the first place is occupied by cardiovascular and cancer diseases, and the third the injury rate. A high level of population mortality is promoted by high alcohol consumption and air pollution.

The levels of morbidity and disability in the cities are closely related to air pollution. This applies for various diseases of the circulatory system, congenital heart anomalies in children, pathologies of the gastrointestinal tract, diseases of the endocrine system, etc. Estimations have shown that in Ulan-Ude, air pollution causes 12% of the total incidence; in Irkutsk, about 40% of children have reduced adaptive capacity due to air pollution (State Report). The population health in the cities of the Baikal region has a number of common features, determined by climatic conditions, atmospheric air pollution, the socio-economic situation, and the state of health. The existing natural differences associated, in particular, with the quality of natural drinking water, are largely offset by the organization of water supply and preventive sanitary and epidemiological measures. Morbidity indicators for many diseases in the cities are higher than in rural areas. However, the life expectancy is an integral indicator of the health in the Baikal region, and after 2007 it became higher than the average regional values in the three regions.

5   Conclusions

Environmental pollution is the chief cause for emergence and formation of persistent foci of anthropogenic preconditions of diseases. The main sources of pollution of the urban environment are the motor transport and industrial enterprises.

A separate group of problem is related to household and other wastes as well as to their disposal. According to the life expectancy, the Baikal region is characterized by lower (by about four years) values against the Russian Federation as a whole. To improve the water supply system for the population and upgrade the water resource quality requires observance of the sanitary norms and a strengthening of the control over the waste water treatment and discharges, an intensification of measures for reconstruction and modernization of the water abstraction and treatment facilities, and the construction of new water supply networks. Water supply of large cities and industrial centers must rely on several water supply sources. Thus, for Irkutsk, Angarsk and neighboring settlements it is advisable to develop an integral water supply system to combine the surface and underground sources in order to reduce the factor of anthropogenic influence on water supply of the population as well as the risk of emergency situations. Air and soil pollution was recorded in large industrial complexes and neighboring settlements. The maps generated in this study can provide a basis for the prevention and avoidance of hazardous ecological situations in the region, the organization of environmental measures and governance of the biochemical environment for livelihoods of the population.

References

1. Vorobyev, N.V., Emelyanova N.V.: The population distribution systems (Chap. 7). In: Geography of Siberia in the Early 21st Century, vols. 6 and 3, pp. 156–176. Economy and Population, Geo, Novosibirsk (2014). (in Russian)
2. Vorobyev, N.V., Emelyanova, N.V., Valeeva, O.V.: Population distribution and dynamics in the central ecological zone of the Baikal natural territory. Geogr. Prir. Resur. (5), 168–178 (2016). (in Russian)
3. Bezrukov, L.A, Aleksandrov, E.A., Vorobyev, N.V, Gales, DA, Grigoryeva, M.A, Dugarova, G.B., Emelyanova, N.V., Zabortseva, T.I., Ippolitova, NA, Ishmuratov, BM, Kuklina, V.V, Leshchenko, S.V, Saraev, V.G., Turkina, N.G, Khavina, L.A., Chelembeeva, A.A.: Assessing the modern factors for urban development and urbanization changes. In: Korytny, L.M., Vorobyev, N.V. (eds.) Geo, Novosibirsk. (in Russian)
4. Gagarinova, O.V.: Landscape-hydrological mapping in research the Lake Baikal basin. Geodeziya i Kartografiya **8**, 14–19 (2016). (in Russian)
5. State Report "On the state of sanitary-epidemiological well-being of the population in Zabaikalkii krai in 2016", Chita, Administration of the Federal Service of Supervision in the Sphere of the Protection of the Rights of Consumers and Human Well-being for Zabaikalskii Krai, 224 p. (2017). (in Russian)
6. State Report "On the state of sanitary-epidemiological well-being of the population in Irkutsk oblast in 2016", Irkutsk, Administration of the Federal Service of Supervision in the Sphere of the Protection of the Rights of Consumers and Human Well-being for Irkutsk Oblast, 275 p. (2017). (in Russian)
7. State Report "On the state of sanitary-epidemiological well-being of the population in the Republic of Buryatia in 2016", Ulan-Ude, Administration of the Federal Service of Supervision in the Sphere of the Protection of the Rights of Consumers and Human Well-being for the Republic of Buryatia, 201 p. (2017). (in Russian)
8. Belozertseva, I.A.: Impact on soils (Chap. 4.3). In: Nature Management in Siberia, A Series of Monographs "Geography of Siberia in the Early 21st Century, vols. 6 and 4, pp. 244–257, Geo, Novosibirsk (2014). (in Russian)

Author Index

A

Abakumov, Alexander I., 73
Ananin, Alexander, 174
Ananina, Tatiana, 174
Ananjeva, Natalia, 57
Andrianova, A. V., 125
Antonov, Igor A., 8
Artemov, Igor, 42
Avramenko, Yurii V., 116, 169

B

Baisheva, E. Z., 80
Bakirova, Rafilya T., 1
Bavrina, Alina, 131
Belozertseva, Irina A., 187, 193
Biktashev, T. U., 80
Bukharova, Evgeniya, 174
Bukin, U. S., 66
Bychkov, Igor V., 116

C

Chepinoga, Victor, 107
Cherkasin, Evgeny A., 96, 151

D

Denisova, Anna, 131

E

Emelyanova, Natalia V., 116, 193
Erbajeva, M. A., 144

F

Fadeev, N. B., 14
Fedorov, Nikolay I., 1, 80
Fedorov, Roman K., 116, 169, 193
Fedoseev, Victor, 131
Feoktistov, D., 159
Firsova, A. D., 169

G

Gachenko, Andrey S., 181
Gagarinova, Olga V., 193
Gatilova, Evgeniya, 22
Gnutikov, Aleksander, 107
Grabarnik, Pavel, 48

H

Han, Irina, 22
Hmelnov, Alexei E., 181

I

Ivanova, Natalya, 48

K

Kavelenova, Lyudmila, 131
Kazanovsky, Sergey, 42
Khalikov, Roman, 57
Korchikov, Eugeny, 131
Kovtonyuk, Nataliya, 22
Kurkov, V. M., 14
Kuzovenko, Oksana, 131

© Springer Nature Switzerland AG 2019
I. Bychkov and V. Voronin (Eds.): *Information Technologies in the Research of Biodiversity*, SPEES, pp. 201–202, 2019.
https://doi.org/10.1007/978-3-030-11720-7

L
Leonov, Michael V., 28
Lobanov, Andrey, 57
Lopatina, Darya N., 187

M
Malkov, Fedor, 151
Mikhailenko, Oksana I., 1
Morozov, Alexey, 151

O
Olonova, M., 159
Orlov, Mikhail, 86
Ostroumova, Tatiana A., 28

P
Pak, Svetlana Ya, 73
Pavlichenko, Vasiliy, 107
Pimenov, Michael G., 28
Pisarenko, Olga, 42
Pishchik, V. N., 37
Popova, Anastasia K., 96
Prelovskaya, Ekaterina, 42
Prokhorova, Nataly, 131
Protopopova, Marina, 107
Pugachev, Oleg, 57
Pukhalsky, Y. V., 37

R
Ruzhnikov, Gennady M., 116

S
Safarov, A. S., 66

Semenov, A. M., 37
Shanin, Vladimir, 48
Shashkov, Maxim, 48
Shekhovtsov, Alexey I., 187
Sheludkov, Alexander, 86
Shigarov, Alexei O., 116, 151
Shumilov, Alexander S., 116
Sidelnikov, N. I., 14
Sinev, Sergey, 57
Skrypitsyna, T. N., 14
Smirnov, Igor, 57
Sorokovoi, Andrei A., 116, 193
Sviridova, O. V., 37

T
Terentyeva, Darya, 131

V
Verhozina, Alla V., 116
Verkhozina, E. V., 66
Verkhozina, V. A., 66
Vladimirov, Igor N., 96, 187
Vorobyov, N. I., 37
Voyta, Leonid, 57
Vysokikh, T., 159

Y
Yakubailik, O. E., 125

Z
Zharkikh, Tatyana L., 1
Zhemyakin, S. V., 37

Printed by Printforce, the Netherlands